Historical Ecology and Archaeology in the Galápagos Islands

Society and Ecology in Island and Coastal Archaeology

UNIVERSITY PRESS OF FLORIDA

Florida A&M University, Tallahassee
Florida Atlantic University, Boca Raton
Florida Gulf Coast University, Ft. Myers
Florida International University, Miami
Florida State University, Tallahassee
New College of Florida, Sarasota
University of Central Florida, Orlando
University of Florida, Gainesville
University of North Florida, Jacksonville
University of South Florida, Tampa
University of West Florida, Pensacola

HISTORICAL ECOLOGY AND ARCHAEOLOGY IN THE GALÁPAGOS ISLANDS

A LEGACY OF HUMAN OCCUPATION

PETER W. STAHL, FERNANDO J. ASTUDILLO,
ROSS W. JAMIESON, DIEGO QUIROGA,
AND FLORENCIO DELGADO

Foreword by Victor D. Thompson

University Press of Florida
Gainesville · Tallahassee · Tampa · Boca Raton
Pensacola · Orlando · Miami · Jacksonville · Ft. Myers · Sarasota

Published in the United States of America

25 24 23 22 21 20 6 5 4 3 2 1

The Library of Congress has catalogued the printed edition as follows:
Names: Stahl, Peter W., author. | Astudillo, Fernando J., author. | Jamieson, Ross W. (Ross William), 1966– author. | Quiroga, Diego, author. | Delgado, Florencio., author. | Thompson, Victor D., author of foreword.
Title: Historical ecology and archaeology in the Galápagos Islands : a legacy of human occupation / Peter W. Stahl, Fernando J. Astudillo, Ross W. Jamieson, Diego Quiroga, and Florencio Delgado ; foreword by Victor D. Thompson.
Description: Gainesville : University Press of Florida, [2020] | Series: Society and ecology in island and coastal archaeology | Includes bibliographical references and index. | Summary: The Galápagos Islands are one of the world's premiere nature attractions, home to unique ecosystems widely thought to be untouched and pristine. This volume reveals that the archipelago is not as isolated as many imagine, examining how centuries of human occupation have transformed its landscape.
Identifiers: LCCN 2019020699 (print) | ISBN 9780813066271 (cloth : alk. paper) | ISBN 9780813057385 (pdf)
Subjects: LCSH: Galapagos Islands—Antiquities. | Indians of South America—Galapagos Islands—Antiquities. | Galapagos Islands—Discovery and exploration.
Classification: LCC F3741.G2 S65 2019 (print) | LCC F3741.G2 (ebook) | DDC 986.6/5—dc23
LC record available at https://lccn.loc.gov/2019020699
LC ebook record available at https://lccn.loc.gov/2019980582

The University Press of Florida is the scholarly publishing agency for the State University System of Florida, comprising Florida A&M University, Florida Atlantic University, Florida Gulf Coast University, Florida International University, Florida State University, New College of Florida, University of Central Florida, University of Florida, University of North Florida, University of South Florida, and University of West Florida.

University Press of Florida
2046 NE Waldo Road
Suite 2100
Gainesville, FL 32609
http://upress.ufl.edu

CONTENTS

FIGURES

PLATES

Plates follow page 76.

TABLES

FOREWORD

When most people think of the Galápagos archipelago, they of course envision many things: the giant land tortoises, the otherworldly saltwater iguanas, sea lions—in short, its "untamed nature." Its close association with Charles Darwin has made it world famous for its natural wonders and their role in shaping his ideas regarding evolution. What perhaps most people do not think about are the people, past and present, who inhabited and still reside in this landscape. What Peter W. Stahl, Fernando J. Astudillo, Ross W. Jamieson, Diego Quiroga, and Florencio Delgado do in this volume is bring to light the human story of Galápagos. In doing so, they reveal that far from being an untouched wonder, Galápagos has been very much entangled historically in both a local and a global human story since the first settlers set foot upon its shores.

When reading this book one is reminded of Sidney Mintz's famous book *Sweetness and Power: The Place of Sugar in Modern History*, which tracks the creation of sugar as a global commodity and its transformation of landscapes. Like Mintz's treatment of sugar, Stahl and coauthors peel back the layers of history of the Galápagos to show how this landscape was transformed by humans into what it is today and how its role in a global economy ingrained this world in the minds of people of the modern world. Thus, this volume does for a place what Mintz's book does for a commodity; that is, it centers on how the history of Galápagos came to be created and reveals for readers the islands' early global connections and the inherent contradictions and struggles in that history.

While Charles Darwin made many observations regarding the plant and animal life of the Galápagos, he also described in detail the people living there at the time of his visit. These settlements, as the authors detail in the first two chapters, were part of a longer history of colonization that occurred both before and after Darwin's visit to the islands. In these chapters, they trace from the possible settlements of pre-Columbian peoples, postulated by the Norwegian hero Thor Heyerdahl and others, to the early and later visitations by whalers and pirates, to its modern landscape and current human geography. These specifics provide the reader with a broader base on which to understand the detailed studies that follow in the rest of the book.

Chapters 2 and 3 outline and detail the history and archaeology of Manuel J. Cobos, San Cristóbal, and the Hacienda El Progreso. As one of the first owners, Manuel Cobos will go down in Galápagos history as one of its more infamous characters. As detailed in the volume, Cobos attempted to transform San Cristóbal into a productive sugar plantation. His development, Hacienda El Progreso, faced a number of challenges, both social and ecological, in its attempts to bring Galápagos into the global economy. That the authors choose to center on San Cristóbal and El Progresso is no mere archaeological convenience, but rather the island, and the Hacienda, are key to understanding historical ecology of the island. The reader gets a sense it is this key period that links the Galápagos distant past to its near history, and to the modern world.

One of the more impressive aspects of the volume is the great diversity of data from the authors' excavations and historical work that they weave into the narrative and questions pertaining to El Progreso and Galápagos in general. Few historical archaeology projects, and in fact few archaeological projects that focus on deeper time periods, can boast such an assembly of data. Specifically, their work includes both archaeofaunal and archaeobotanical (both micro and macrobotanical) analyses; the latter of these also includes an examination of wood charcoal that provides important clues to the introduction of tree species and their concomitant use on the island. The authors of course couple these analyses with more traditional and historical methods, which include the analysis of standing architecture and historic photography. All these data allow the authors to speak to the complex introductions and interactions of past humans with plant and animal species—both those that were endemic and those that were invasive to the island.

The penultimate chapter details the material culture of the excavations. What is striking about the observations that the authors draw from these more often than not small objects is the insights they provide into the global connections present at the time for this archipelago in the Pacific. What is apparent is that the global consumerism that was taking hold of much of the world at that time reached Galápagos as well. We also see how race and class also impacted the nature of what was consumed, and there are hints too about often overlooked persons in the archaeological record—the children of El Progreso. In all, Galápagos was a place still linked to the larger world, and depending on who you were in this community, there was a wealth of consumer goods to choose from despite the seeming isolation of these islands.

As the authors point out in their conclusion to the volume, Galápagos, in varying degrees, "has always been a globalized place." This is one of the key theses of the work and in many ways, it is from this perspective that the work in a nuanced way challenges our very ideas of "nature." Furthermore, the authors use their work in the previous chapters to explore the future of Galápagos. The character of human settlement on San Cristóbal and the other islands has varied over time, from whaling waystation, colonizing sugar plantation, American Tuna Fleet stop, and wealthy tourism funneled by the construction of the Panama Canal. The archaeology and history of these islands truly describe its links to the larger world. In the end, this work forces us to confront the collective idea of the Galápagos as a natural place and realize that these connections continue today. As the world grows smaller and more connected, the future of the Galápagos landscapes become more uncertain.

Victor D. Thompson
Series Editor

ACKNOWLEDGMENTS

The Historical Ecology of the Galápagos Islands project began more than eight years ago when Florencio Delgado invited Peter Stahl to collaborate with him in an investigation of the historic Hacienda El Progreso on San Cristóbal Island. With funds awarded to Delgado through a Galápagos Academic Institute for the Arts and Sciences (GAIAS) Collaboration Grant from the Universidad San Francisco de Quito (USFQ), he and Stahl traveled to the island in 2012 for a 10 day exploratory trip while both were working together on an archaeological project in Manabí Province. We thank USFQ and its office of the Dean of Research for their generous financial support, which enabled us to visit El Progreso and begin our plans for a larger study. Stahl subsequently obtained funding in 2014 through a Partnership Development Grant (890-2013-0013) from the Social Sciences and Humanities Research Council of Canada. We are grateful for this multiyear award which enabled us to partner the University of Victoria (UVIC) with USFQ, Simon Fraser University (SFU), GAIAS, the Galápagos Science Center (GSC), and the Junta Parroquial, Gobierno Autónomo Descentralizado Parroquial El Progreso.

With authorization from the Instituto Nacional de Patrimonio Cultural (INPC) and permission from the Parque Nacional Galápagos (PNG), we were able to undertake five seasons of investigation and analysis between 2014 and 2018. We thank both institutions for facilitating our project, and acknowledge the enthusiastic support of Marcos Labrada, director of the Special Service Center for Zone 4 of the INPC, and Joaquín Moscoso, its executive director. None of this would have been possible without the active support of the Junta Parroquial, and the kindness and generosity of the El Progreso community, who welcomed us as fellow, albeit transient, residents. We are especially grateful to our good friend Paulina Cango, Presidenta del Gubierno Parroquial El Progreso, for her thoughtful leadership, kindness, and undying support. Our heartfelt welcome in El Progreso was expressed through the benevolence and openness of many in the community, especially, but not limited to members of the Becerra, Jiménez, Libiapoma, and Sumbana families, and individually Jimy (Pata Larga) Becerra, Juan Carlos Becerra, Luís Chango, Roxana Culqui, Eli Libiapoma, Mercedes Medina, and GNP guide Jeffress Málaga. Eli, in particular, granted access to his property while he patiently continued working in his shop directly above the main midden where our excavations simultaneously created extra space for his courtyard. We reserve special mention for our colleague and pana, Edy Bismark Becerra. Edy's larger-than-life personality, unwavering good cheer, and unrestrained knowledge were crucial for every phase of the project. We are both glad and fortunate to have met Edy on the very first day of our arrival, and we continue to benefit from his sincere interest in the history and future of his community.

Various students and colleagues assisted in the project over the course of five field seasons. We extend our thanks to Celín Astudillo, Kirby Booker, Francis Castillo, Martín Delgado, María Isabel Guevara, Jescima Janson, Andrea López, Maeve LeDuc, Tamia Maldonado, Alex Masaquiza, Augusto Oyuela-Caycedo, Peter Raskovsky, Miranda Riou-Green, Geovany Sarigu, Eric Simons, Tanya Taggert-Hodge, Josefina Vásquez, Brock Wiederich, and María José Yepez. We are particularly indebted to Tanya who, with the assistance of Celín, capably produced the repeat photography. Aerial and terrestrial LiDAR mapping was undertaken through the Hyperspectral-LiDAR Research Lab in the Department of Geography (UVIC) under the direction of Olaf Niemann. We thank Olaf, who generously facilitated our research, and especially Georgia Clyde, Cydne Potter, and Roger Stephen for accompanying us to San Cristóbal in 2018 and skillfully undertaking the study. For their assistance in realizing the success of the mapping study, we are also deeply indebted to the efforts of Gonzalo Rivas-Torres and Cristina Vaca (USFQ).

We were very fortunate to have received crucial assistance from the GSC and GAIAS at Playa Mann, Puerto Baquerizo Moreno, and are deeply grateful for the kind and generous support of GSC co-directors Steve Walsh (University of North Carolina–Chapel Hill) and Carlos Mena (USFQ), and the assistance of Juan Pablo Muñoz, Sylvia Sotamba, Luís Tasipanta, and Leandro Vaca. At GAIAS, we acknowledge the assistance of Cecibel Narvaéz and Máximo Ochoa in Playa Mann, and Sofía Tacle in Quito. At UVIC we were ably assisted by Jindra Belanger, Ute Mueller, and Cathy Rzeplinski (Anthropology); we also acknowledge the support of Ann Stahl (Anthropology) and Peter Keller (School of Social Sciences), and we owe a deep debt of gratitude for the crucial help of Rosemary Ommer and Terri Myhr (Research Services). We are happy to acknowledge the expertise and cheerful help of Jessica Fitterer (Geography) and her protégés Ben McGrath and Rylee Harlos for their mapping skills. Interlibrary loan services at UVIC Libraries graciously accommodated our numerous requests. At USFQ we acknowledge the support of Dean Carmen Fernández-Salvador, and the accounting acumen of Juan-Carlos Chanaba. At SFU, project archaeobotanical research benefited from the assistance and comments of Catherine D'Andrea, Francesco Berna, Dana Leposky, Shannon Wood, Peter Locher (Archaeology), and Sarah Walshah (History). Dongya Yang and Thomas Royle of the SFU Ancient DNA Facility lent their expertise to the zooarchaeological study of marine turtle specimens. Madelyn Percy (Geology, UNC–Chapel Hill) kindly described the midden soil contexts for us. For bringing this book to fruition, we are indebted to Victor Thompson (University of Georgia), Meredith Morris-Babb of the University Press of Florida, and to the anonymous reviewers of the book manuscript. We are of course responsible for its contents.

1

Introduction

For many tourists, their first glance of Galápagos appears abruptly at the end of a one-and-a-half-hour commercial flight from Guayaquil's José Joaquín de Olmedo international airport. The sight is breathtaking, even to the seasoned visitor. As the plane begins its descent through the clouds, the greens and browns of San Cristóbal Island, which suddenly come into view, contrast strikingly against the emerald and blue waters of the Pacific Ocean. The moment is often greeted with hushed gasps as passengers peer through their windows to catch a glimpse of a once-in-a-lifetime sight. As the plane maneuvers into its approach, details of the island become clearer, and conversation often turns to one of surprised observation that people actually live here. First-time visitors are confronted with this realization as the jetliner taxis into the newly refurbished San Cristóbal airport, only walking distance from the provincial capital of Puerto Baquerizo Moreno.

After entering the Galápagos National Park through the airport arrivals building, visitors encounter a bustling curbside pickup area framed against the backdrop of a heavily quarried hill, which serves as the islands' principal source of building material. The quick ride into town usually terminates at the passenger dock, where luggage is unloaded for transfer via a smaller launch to the tour boat waiting at its berth farther out in Wreck Bay. Some passengers immediately embark on their planned cruise around the islands; others often spend the few preboarding hours at Playa Mann, a short walk to the other end of the bay. For prepackaged tours originating on San Cristóbal, the island adventure ends in reverse order as tourists are whisked off to the airport, after which the process is repeated with the anticipated arrival of yet another planeload of visitors.

The average tourist to Galápagos spends little time on the island, and only the most adventurous travel up the road into the interior highlands. More often than not these visitors come to see the crater lake at El Junco, maybe visit La Galapaguera, where tortoises are raised for restocking on various islands, or have their rented bicycles transported in hired pickup truck taxis so they can leisurely bike downhill to the coast. In either case, they must pass through the small village of El Progreso, usually keeping to the southern branch of the road which avoids the center of town. The community rarely sees any tourists stopping on their way through the village, a constant reminder that the island's principal sources of interest and revenue are centered on the coast.

During this brief foray into the island's interior, the adventurous tourists pass through a heavily vegetated landscape dotted with entrances to family farms and the occasional grazing cow. If they proceed further into the highlands they will pass open fields, an occasional building, three large wind turbines, and the parking lot for El Junco. Eventually, they descend through more vegetation and farm entrances, the small hamlet of Cerro Verde, and La Galapaguera, before reaching the road's terminus at the parking area for Puerto Chino.

This quick, roughly 30 km trip takes the visitor through the southwestern tip of the island and one of the few Zones of Special Use (ZUE, Zonas de Uso Especial) in the Galápagos National Park (PNG, Parque Nacional de Galápagos). These zones, which occupy approximately 3.3% of the PNG's terrestrial area, are located primarily in the highlands of the four populated islands: San Cristóbal, Santa Cruz, Floreana, and southern Isabela (Figure 1). The ZUEs were recognized by the PNG in 1996 as areas where the park's management plan permits uses contradictory to the primary objectives of protected areas, specifically ecosystem conservation and recreational activities. The PNG recognizes these areas as Human Space, which includes Urban Zones of previously privatized areas on the inhabited islands dedicated to commerce, industry, and residence, and Rural Zones in the highlands of populated islands where the best agricultural soils are located. On San Cristóbal, some 93% of the island's original humid habitat is an agricultural zone (INGALA 2005).[1]

Surrounded by park, the ZUE on San Cristóbal can be distinguished from PNG lands by a sharp boundary that is clearly visible from various highland viewpoints where the duller browns, mauves, and greens at lower elevations contrast abruptly with the often canopied verdant greens

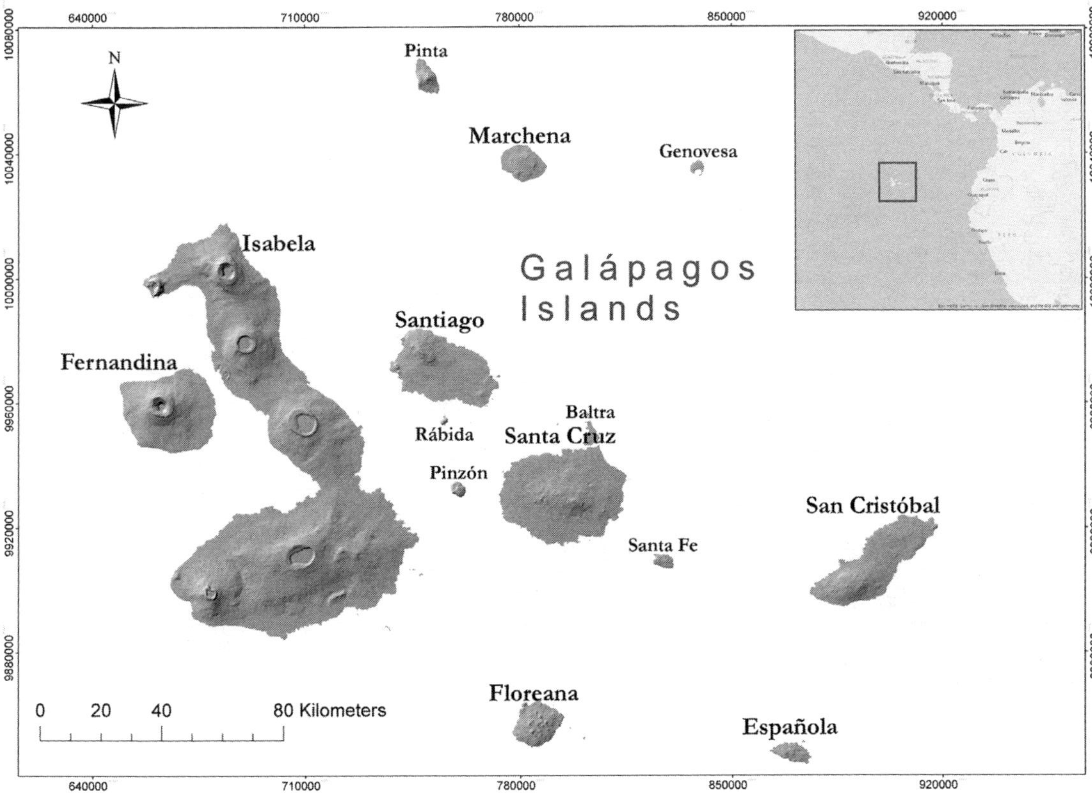

Figure 1. Map of the major islands making up the Galápagos archipelago and their relationship to the South American mainland (*inset*). Map Datum: UTM 15S WGS 1984, created by J. Ben McGrath.

of the higher-elevation agricultural zone. This zone, and its counterparts on the other islands, is the most heavily impacted area of exotic invasive alien organisms in the PNG.[2] On San Cristóbal it survives today as the legacy of earlier human colonization on the island, particularly the historic Hacienda El Progreso, which endures today in and around its contemporary namesake.

Tourists don't travel to Galápagos to experience El Progreso or ZUEs. They come to experience the natural beauty of a world-renowned nature destination. The island archipelago is regularly included in every list of top vacation spots for observing natural wonders. Indeed, since its earliest beginnings, the park was conceived as a self-funded nature conservation area sustained by income from boat-based tourism. Limiting the numbers of boats, and controlling access of park visitors to terrestrial habitats, served the dual purpose of mitigating potential disturbance while simultaneously raising the revenue required to regulate restricted access. Since 1978, Galápagos has been deemed a UNESCO World Heritage site for meeting as many as four natural selection criteria. It contains superlative phenomena or exceptional beauty, important or significant habitats for biodiversity conservation, and represents both major steps in earth's history and significant ongoing ecological or biological processes. UNESCO considers Galápagos a "living museum and showcase of evolution."[3]

Since Charles Darwin's landmark visit to four of the islands in 1835, Galápagos has assumed an iconic position in the study of evolutionary principles.[4] Composed of true oceanic islands with unique insular endemics, and effectively isolated from its closest continental landmass, Galápagos is geologically young enough to reveal evolutionary change, and it includes sufficient variety to study multiple isolation events in a pristine setting relatively isolated from humans (Valle and Parker 2012); however, it is the pristine attribution of the archipelago that is in danger, certainly as a corollary of its increased fame. The perception of Galápagos as a people-free "natural laboratory" for understanding evolution appeals to an educated and prosperous clientele and fuels a lucrative multimillion-dollar tourist business. The latter contributes to a paradox in which its appeal as a pristine natural laboratory creates the very conditions that undermine its appeal (Quiroga 2009).

The continuous growth of tourism, including both traditional boat-based tourism and increasingly land-based coastal tourism, is attracting record numbers of human visitors to the islands. While it affords economic opportunities for Ecuadorians, the lucrative industry simultaneously places strains on existing infrastructure. Conservation priorities strongly correlate increased human visitation with the introduction of alien organisms that can become naturalized and invasive with or without intentional human intervention (Pyšek et al. 2004). The proliferation of invasive alien species represents the single largest threat to terrestrial ecosystems throughout the PNG today (Atkinson et al. 2012; Carrier et al. 2011; Gardener et al. 2010).

El Progreso, ZUEs, and most references to historic associations between humans and Galápagos ecosystems are usually excluded from the narrative of nature because they depart from the image of the pristine. Invasion by alien organisms has, nevertheless, always been a part of Galápagos history; the archipelago's unique insular organisms are themselves endemic legacies of invasions that arrived in these pelagic islands over the course of geological time from continental and island sources around the Pacific (Grehan 2001); however, this does not mitigate the severe problems that naturalized and invasive aliens pose for endemic ecosystems. Galápagos is particularly susceptible to invasion owing to its small scale and the relative poverty of its endemic species, which having evolved in isolation, are often competitively inferior to their alien counterparts (Whittaker and Fernández-Palacios 2007: 320). In various parts of the archipelago the extent of invasion, in numbers and kind, is alarming. Conservation's ultimate goal is to reestablish native biodiversity and ecosystem function (Atkinson et al. 2012: 183). Eradication efforts, many of them high profile, have been under way.[5]

Unlike many other oceanic islands, the Galápagos archipelago may have been relatively isolated from alien invasions because of its remote location away from international trade routes, and its later colonization by humans (Whittaker and Fernández-Palacios 2007: 320–321). Although alien introductions are recognized from as early as the seventeenth century, their major impetus is usually considered late and coinciding with the steady rise of human visitation associated with tourism after the 1960s. Up to this point, the number of alien plants is believed to have increased slowly, followed by a quick rise, mostly for cultivation. An abrupt increase subsequently occurred in the 1980s, with growth proceeding exponentially thereafter, especially on the inhabited islands (Mauchamp 1997; also Tye 2001; Tye at al. 2002). A linear increase of alien invasives is typical for most islands, and the apparent exponential pattern in Galápagos is debatable. It may have occurred in a stepwise pattern, with an early stage up to 1911, followed by a second stage after the 1960s, and an explosion after 1987 (Tye 2006). In any case, "there is continuing uncertainty about the real human-mediated introduction rate of plants to the archipelago, which is of conservation concern" (Tye 2006: 203).

A precise chronology for ordering the number and rate of alien introductions into the Galápagos Islands will always be elusive; however, the timing and circumstances surrounding human interactions with island ecosystems in Galápagos is more lucid than for most island contexts. Galápagos' relatively late discovery by humans, probably beginning no earlier than the first half of the sixteenth century, is followed by a frequently recorded series of historical encounters. By combining historic documentation with the legacy of preserved traces in the archaeological record, the timing and circumstances surrounding alien introductions can be further illuminated, thereby providing increased temporal resolution and historic context to the establishment and longer-term development of legacy landscapes throughout the archipelago.

Historical Ecology, Conservation, and Archaeology in the Galápagos Islands

The image of a people-free natural laboratory for understanding evolution is essential to the existence of a lucrative Galápagos-based ecotourism industry, whose priorities are strongly inclined to understate or altogether omit cultural landscapes, contemporary or historic. The various efforts that attempt to restore Galápagos ecosystems to precolonized conditions engage different political, economic, and scientific stakeholders in a scientific endeavor guided by evolutionary theory to shape island ecosystems according to the demands of ecotouristic expectations, which are based upon Western idealizations of a "wild" nature that still endures in one of its last remaining refuges (Hennessey and Cleary 2011; Quiroga 2009).

Nevertheless, increased human intervention has intentionally and unintentionally transformed previously isolated terrestrial Galápagos ecosystems into "novel" or "emerging" ecosystems with structures composed of new species arrangements defined by their novelty, anthropic origin, and subsequent endurance in the absence of humans (Hobbs et al. 2006, 2013; Morse et al. 2014). Despite years of intensive effort and enormous capital outlay, battles against invasive aliens and efforts to restore terrestrial ecosystems to prior conditions have met with uneven, and at times limited, success in various parts of Galápagos (Vince 2011). Some argue that restoration has offered unrealistic expectations and false promises, particularly as former ecosystems will never return. Novelty should neither be denigrated nor ignored; rather it is to be studied and embraced, particularly where stable ecosystems have become critical for maintaining biodiversity and for providing services to human stakeholders.[6] Minimally, the insertion of novelty into conservation management can also guide decisions regarding appropriate intervention and the effective use of limited resources (e.g., Bridgewater and Higgs 2011; Collier and Devitt 2016; Hobbs et al. 2011, 2013, 2014; Lugo 2009; Quiroga and Rivas 2017).

Opponents of this perspective on novelty consider it unwise to abandon restoration because it is often a feasible management approach that can also provide benefits. Moreover, novelty is not necessarily inevitable as many ecosystems remain intact, and it is difficult to judge when a threshold of irreversibility, the point after which ecosystems tip to novelty, has been reached. Novel or hybrid ecosystems that contain "junk species" are not equivalent counterparts to their predecessors. The acceptance of novelty is an entry point that can lead conservation down a slippery slope to ruin, in addition to conveniently providing economically motivated development with a "license to trash" or a "get out of jail free card" (Doak et al. 2014; Hobbs et al. 2014; Murcia et al. 2014; Standish et al. 2013; Vince 2011).

Escalating novelty and differing approaches to its management lie at the core of a contentious and polarizing debate between supporters of a new Conservation Science and adherents to a more traditional Conservation Biology. The former reject the notion of restoring pristine Holocene ecosystems and advocate a resilient nature that includes humans and their actions. They support a form of conservation management that takes into consideration ecosystem services benefiting the most humans, particularly those disenfranchised by restoration schemes. Their detractors dismiss the numerous criticisms leveled against traditional Conservation Biology by Conservation Science, claiming that the latter is ethically rather than scientifically based. When Conservation Science emphasizes ecosystem services for human profit, it diminishes the importance of preserving biodiversity and perhaps does not deserve to be considered conservation (e.g., Doak et al. 2014; Hunter et al. 2014; Kareiva et al. 2007, 2011; Kareiva and Marvier 2012; Soulé 2013).

Investigations of the archaeological record can be productively combined with existing documentary sources through the perspective of Historical Ecology (Balée 1998, 2006; Balée and Erickson 2006) to explore issues surrounding the origin, temporal and spatial development, and historic context of enduring humanized landscapes throughout the Galápagos Islands. Humans and the biospheres they inhabit are interconnected in an intimate and historically contingent dialogue. Their cultural expertise regularly places humans in keystone roles (Mills et al. 1993) within the landscapes they assemble wherever they reside. Historic circumstances involving the social, economic, political, and religious contexts within which this relationship develops often govern the nature and extent to which it affects local biological diversity.

Although humans have relentlessly colonized much of the world's land surface since their appearance as a species, Galápagos provides an effective illustration of an initial peopling event owing to its isolation and relatively recent colonization. It can provide a rare glimpse of humans encountering uninhabited environments and entering into a transformative relationship that initiated the creation of humanized landscapes. Who were the humans that colonized previously uninhabited space? What were the various social, economic, political, religious, and psychological circumstances that historically motivated their colonizing efforts? Although the Galápagos Islands were frequently attractive to colonists because of their isolation and imagined pristine qualities, even their remote Pacific location could not insulate them from global events whose reach eventually extended into the most inaccessible areas.

Upon arrival, where did early colonizers locate their settlements? What were the absolute necessities for short-term survival? Water was an immediate concern, and in Galápagos, lower elevations and beaches were particularly lethal areas for terrestrially bound human visitors. What was needed to ensure longer-term survival? Highland areas with regular water sources and soils that could sustain agriculture were imperative. Islands lacking these linked resources had to be avoided. What did early colonizers bring with them to increase their chances of longer-term self-sufficiency and survival? They imported acquired knowledge, tools, materials, and familiar plants and animals required to re-create a relinquished world, which they transported as a mental template. The reproductive propagules of seeds, cuttings, and breeding pairs simultaneously became the pioneer organisms required for sustaining human life and the ancestral stock for future invasive aliens. They composed the "portmanteau biota" of fellow life forms, an extended family of organisms adapted to and by humans, which was crucial to their successes and subsequent failures as they colonized the globe (Crosby 1986).

Successful farming requires energy, luck, and time. Even the fastest-growing plants could not meet the immediate needs of colonists. Survival initially required the exploitation of local resources, certainly as imported necessities were both limited and of vital consequence for future success. From the outset, human pioneers regularly depended on the legacy of feral resources abandoned by previous colonization during their historical encounters with Galápagos. Human settlers repeatedly sought out the landscapes left behind by previous settlement. These contained the feral resources and preserved infrastructure upon which to build anew, and they offered the logistical choices that originally made them suitable locations for colonization. Ironically, early Galápagos settlers frantically searched for any kind of fencing material to keep out of their growing gardens the naturalized feral stock they had earlier depended on for their survival. Despite their often stated wish to colonize these remote lands in order to live in relative seclusion, early colonizers continued to rely on vital connections with the world they had forsaken. Exchange of local production for needed replacements, but especially news and basic companionship, were eagerly anticipated from abroad. These persistent links that humans maintained to interconnect themselves continued to expand through time and space to envelop even the remotest communities in broadening global networks.

The study of early human colonization in Galápagos can contribute to a wide range of intersecting interests in ecology, archaeology, and history. It supplies a much needed historical dimension and offers a humanized context for our understanding of vital issues associated with management and conservation in an iconic global nature park. By adding a social background and providing examples of historical contingency for the study of initial human colonization, it helps us interpret the scale, extent, and speed of early colonization, the various ways in which humans and the new environments they encountered interacted with each other, and it offers a comparative study for how we consider the form, meaning, and aftermath of colonization in various world areas (e.g., Derevianko et al. 2005; Hannon et al. 2001; Mellars 2006; Mulvaney and Kamminga 1999; O'Connell and Allen 2015; Vésteinsson and McGovern 2012; Waters and

Stafford 2012). Understandably, the most instructive examples of human colonization are often provided by island contexts, and Galápagos contributes an important, hitherto unexplored, archaeological case study of human impact on pelagic island ecosystems in the Pacific Ocean (e.g., Burney and Burney 2007; Gibbons 2000; Hunt and Lipo 2009; Kennett et al. 2005; Kirch 2000; Mann et al. 2008).

The study of Galápagos colonization also contributes to our understanding of the ways in which human intervention affects local ecosystems, including the impact of introduced technologies and organisms, the development of humanized landscapes, and their aftermath. What organisms were intentionally introduced and what motivated their selection? What organisms were unintentionally introduced and what may have been the mechanisms behind their unintended introduction? How did these organisms impact local ecosystems, and what did they share in common with introduced organisms elsewhere? What native resources were exploited, translocated, managed, or selectively encouraged and which ones were removed, isolated, depressed, or became extinct? What were the effects of human landscape technology on local abiotic conditions, including soils and hydrology? What was the aftermath of human intervention on the structure and resiliency of local ecosystems, and how does it compare with other world areas? Galápagos adds historical and archaeological documentation for recording the effects of human colonization on local ecologies in various world areas, and especially on islands (e.g., Anderson 2002; Athens et al. 2002; Braje and Erlanson 2013; Briggs 2011; Burney and Flannery 2005; Curchamp et al. 2003; Grayson 2001; Hofman and Rick 2018; Kirch 2005; Martin and Steadman 1999; Reitz 2017; Steadman 2006).

Galápagos Sugar Imperium

The focus of this collaborative study is the human history of Galápagos and specifically the historic Hacienda El Progreso, perhaps the most ambitious of a number of early Republican period colonization efforts in the archipelago. From its modest beginnings in the 1860s to its founder's dramatic assassination in 1904, El Progreso rapidly became the largest and most ambitious island colony in Ecuador's remotest western frontier. Carrying on in various forms for another sixty years, its legacy survives today throughout the southwestern end of San Cristóbal Island. The historic hacienda targeted many of the major naturalized alien organisms that plague the PNG today, but at its very core lay the industrial-scale production of sugar, whose extraction and refinement created one of the most demanding and labor intensive of all human crops.

Since its early domestication in tropical Southeast Asia, sugar undertook its relentless western march through Asia, the Mediterranean, and Africa, to the eastern Atlantic islands of São Tomé, Cabo Verde, and the Canaries, before undertaking its extensive and transformative colonization of the New World (Abbott 2008; Galloway 1989; Mintz 1985). It eventually made its way to the extreme western coast of South America (Gonzales 1985), after having been established throughout many of the warmest regions of Europe's far-flung empire. Hacienda El Progreso represents a relatively late extrusion of sugar production into one of its remotest outposts in the eastern Pacific, almost 1,000 km from the Ecuadorian coast. Although modeled on extant templates of Spanish colonial production (e.g., Barrett 1970; Hazard 1871; Moreno Fraginals 1976; Tezanos Toral 2015), it nevertheless also provides an example of sugar's transition to modern techniques and methods.

The following chapters recount the history of humans in Galápagos. A particular emphasis is placed on the interconnections between anthropic transformations of distinctive island landscapes, the increasing integration of Galápagos into an expanding global network of human interests from different continents and hemispheres, and the changing perceptions of nature held by humans. Chapter 2 begins with the uninhabited archipelago's accidental discovery by Bishop Tomás de Berlanga and ends with post-World War II interests in conservation and tourism. It emphasizes the growing and shifting interrelationship between humans and island ecosystems through time, as Galápagos was increasingly frequented by seventeenth-century buccaneers and eighteenth-century whalers and sealers. It continues into early nineteenth-century annexation by Ecuador and Republican-era colonization. The twentieth-century completion of a Panama Canal increased ease of access that ushered in succeeding visits by researchers and adventurers, some who stopped temporarily and others whose descendants remain to this day.

Chapter 3 focuses on the island of San Cristóbal, specifically its southernmost end, site of the historic Hacienda El Progreso and current home to its legacy of the same name. It tracks the development of the hacienda through the life of its founder, Manuel Julián Cobos, from his early business in coastal Ecuador and establishment of a workforce on the island in 1866, through his sojourn in Baja California, to his return to San Cristóbal in 1879. It follows the subsequent quarter century of developments at Hacienda El Progreso from its humble beginnings, through its apogee as an industrial-scale, export-focused sugar plantation and ranch with hundreds of employees, to Cobos's death at the hands of his workers in 1904. It traces subsequent developments associated with changing ownership and management, to its eventual attempted colonization by disillusioned American expatriates in the 1960s. It ends with a brief description of current human geography in Galápagos and includes a discussion of landscape transformations and naturalized alien introductions in island ecosystems.

Chapter 4 details archaeological studies of historic Hacienda El Progreso, which after a brief exploratory trip to the island in 2012, were carried out over five consecutive seasons between 2014 and 2018. After describing the areas of archaeological activity around the town site and surrounding landscape of El Progreso, the chapter details the surviving legacy of historic sugar production. Zooarchaeological and archaeobotanical studies present the preserved evidence for exploitation of exotic and wild endemic animals and for exotic crop plants, woody taxa, and the hacienda's impact on native vegetation. The chapter concludes with a series of repeat photography images, juxtaposing contemporary imagery on precisely aligned historic photos that visually document historic landscape transformations at various points in time.

Chapter 5 continues with a presentation of the preserved material cultural assemblage recovered in excavations. It examines technologies of control and elite consumption, with Cobos and the historic hacienda serving as a nexus of commodity introductions into remote areas of the Pacific. This was undertaken as Latin America participated through export of commodities and import of manufactured goods in the global markets of an "Export Age" between 1870 and 1930, defined by access to credit and demand for tropical products. Our study concludes in Chapter 6 with closing remarks about culture inside a nature park. It considers the creation of the park and biosphere in an "Age of Leisure," the impact of tourism, and the future for El Progreso and island residents as stakeholders in a conservation ethos.

2

Humans Encounter Galápagos

Running dangerously low on water after ocean currents had swept his ship far into the Pacific, Tomás de Berlanga, the fourth Bishop of Panamá, fortunately sighted an island on March 10, 1535. In his subsequent letter to King Charles V, Fray Berlanga (1884[1535]) describes a worthless landscape of slag in which neither a bushel of corn could be sown nor even a little grass raised. Filled with very large rocks, it seemed as if God had once showered it with stones. He briefly mentions sea lions, iguanas, marine turtles, and very large tortoises, or galápagos, the latter by which his inadvertent discovery would come to be referred. Over the years, this isolated collection of volcanic peaks emerging above the ocean surface almost 1,000 km from the South American coast has received many names,[1] at least some of which reveal bleak human appraisals like *World's End, The Rock,* or *Clinker Islands*. Throughout their human history, Galápagos were frequently disparaged as less than useful real estate.[2]

Today, our perceptions of Galápagos can be gauged by what we write about them. Thousands of books and scholarly articles discussing biology and conservation are readily available, whereas far less is devoted to their human history. The islands' prominent position in scientific thought, coupled with their enviable position as one of the world's premier nature tourism destinations has simultaneously heightened our attention and boosted their attraction while diminishing a shared and complex interaction with humans over the course of centuries. When acknowledged, human history throughout the Galápagos archipelago is usually cast in terms of a negative legacy to be extirpated through restorative efforts.

Nevertheless, this long, complex, colorful, and at times depressing history is available in a number of publications.[3]

This chapter summarizes major events that occurred throughout the almost 500-year relationship between Galápagos and humans. Specific attention is focused on how the latter's changing interests contributed to ecosystemic and particularly landscape transformation up to the end of the Second World War. The following chapter details the late nineteenth century apogee of these interests, expressed in industrial-scale extraction and export on San Cristóbal. The framework of Historical Ecology (Balée 1998) guides this exploration; of particular importance is the historically contextualized nature of these changes as direct human activity interacted with relatively recently encountered island ecosystems, and how this interconnected relationship shifted through time within the context of changing political and economic circumstances.

Possible Pre-Columbian Encounters

Shortly after Vasco Núñez de Balboa crossed the Isthmus of Panamá in 1513, early reports repeated the promise of rich islands with spices lying further to the west in the South Sea. One example, previously used as evidence for a visit to Galápagos as early as 1516, is found in the third decade of Peter Martyr's *De Orbo Novo*[4] which records the account of troops under the command of Gonzáles de Padajoz. The report, however, mentions herds of wild deer and boar (MacNutt 1912[1530]: 409) which more likely suggests one of Panamá's many offshore islands. Early accounts also probably inspired the widely cited evidence for Incan subjugation of Galápagos by Tupac Yupanqui. Pedro Sarmiento de Gamboa's 1572 *Historia Indica* briefly describes the Inca's expedition to the islands of Avachumbi and Ninachumbi. Acting on accounts of navigators who had arrived from these rich and populated islands on balsas with sails, the Inca equipped a large force of 20,000 men, returning more than nine months later with black people, gold, a chair of brass, and the skin and jawbone of a horse (Markham 1907[1572]). The veracity of these claims is widely discredited for both their implausibility and their authorship. Sarmiento was certainly one of the most peripatetic chroniclers of Spain's new empire, and his own interest in the improbable account led him across the Pacific in 1568 to find Tuvalu (the former Ellice Islands), Guadalcanal, the Marshall Islands, and Wake Island before returning to lower California later in the same year, having missed Galápagos entirely (Clissold 1954).

Nevertheless, the long-distance seafaring capacity of indigenous peoples along the equatorial coast of South America is well known. The accounts in many early chronicles and from later European mariners describe their vessels.[5] Four previous archaeological expeditions have visited Galápagos, primarily to investigate pre-Columbian visitation to the islands lying some 600 nautical miles from the coast,[6] the most famous of which located five sites on three islands with good landing places and available water. All were shallow sites with no visible stratigraphy, and all contained colonial pottery interspersed with imputed pre-Columbian ceramics (Heyerdahl and Skjőlsvold 1956). Multiple lines of preserved evidence have recently been used to reject these and subsequent claims for ancient visitation of the islands (Anderson et al. 2016).

Early Colonial Encounters: Buccaneers and Pirates

Berlanga's accidental discovery of Galápagos in March 1535 ended happily almost a month later when he entered the Bay of Caráquez on Ecuador's central coast. His detour had occurred en route to Peru, where he was ordered to report on conditions in the newly conquered territory while settling a territorial dispute between the conquistadors Francisco Pizarro and Diego de Almagro. Little is known of subsequent visitations by the Spanish, or anyone, during the sixteenth and early seventeenth centuries; however, a series of maps, the first published anonymously in 1561 and later examples with Spanish names,[7] suggest that they had been visited with relative frequency (Latorre 2011).

Fighting on the side of loyalists and pursued by forces allied with Gonzalo Pizarro's rebellion against the Crown's New Laws ensuring the protection and rights of conquered indigenes, Diego de Rivadeneira fled by ship from the southern coast of Peru in 1546. Lacking charts or compass, Rivadeneira and at least a dozen men set out for New Spain in Central America. After 25 days of hardship they encountered mountainous and rocky islands containing large quantities of sea lions, turtles, iguanas and birds, but no water. Surviving on rainwater and marine animals, they eventually reached Guatemala. Two other credible accounts of island visitation were also recorded in 1585 and 1586 (Jiménez 1891).

From early in the colonial experience, Spain had been transferring the wealth of its growing Pacific empire via Caribbean transshipment centers in Vera Cruz, Portobelo, and Cartagena. Designed to maximize the acquisition of precious metals, particularly silver, from Mexico and Peru, the Spanish relied on a preexisting pattern of one-port monopolies for transporting products via escorted convoys to Sevilla, and later to Cadiz. These Caribbean-based Flotas de Indias became a permanent feature in 1561, annually shipping millions of pesos in precious metals alone, and turning the West Indies into "the strategic center of Spain's transatlantic empire" (Stein and Stein 2000: 4). As European competitors began to establish Caribbean bases from which to attack the convoys, Spanish naval operations shifted to protecting their maritime lifeline from the increasing piracy responding to the flow of New World silver (Stein and Stein 2000). Also, after 1565, when the Spanish had discovered a return route from the Philippines to their Pacific transshipment center in Acapulco, the Manila galleons or *naos de China* regularly began to supply a ravenous Chinese market with silver in return for luxury goods in great demand in New Spain (Gerhard 1960).

With the exception of a few privateers, especially the sudden appearance of Sir Francis Drake in 1578, the eastern Pacific remained an exclusively Spanish realm during the sixteenth century.[8] One of Drake's goals was to free John Oxenham, who having crossed the Isthmus in 1575 to attack treasure galleons from the Pearl Islands, was captured and sent to Lima for questioning by the Inquisition. Their early successes, particularly Drake's legendary seizure of the galleon Nuestra Señora de la Concepción or *Cacafuego* aroused interest in South Sea treasures. Later expeditions by Thomas Cavendish in 1587, Richard Hawkins in 1594, and Dutch raiders or *pechelingues* at the close of the century met with varying success (Gerhard 1960).

Three Dutch expeditionary fleets under the command of Speilbergen in 1615, Schapenham in 1624, and Brouwer in 1642 entered Spain's Pacific realm, which otherwise remained relatively calm until the incursion of Caribbean buccaneers in 1680.[9] Pirate bands were making inroads across the Isthmus during the mid-seventeenth century, and in 1671 a force under the command of Henry Morgan sacked Panamá, the Pacific transshipment port to Portobelo, which encouraged buccaneers into the South Sea (Gerhard 1960). In April 1680, a large force of more than 300 men with seven ships under the command of Richard Sawkins crossed Panamá to the Pacific. Notable in this group were Capt. Bartholomew Sharp, William

Dampier, and Lionel Wafer. After commandeering Spanish ships in the Pacific and Sawkins's death, Sharp assumed command, continuing to raid the western coast of South America for ten months (Gerhard 1960). In June 1680, he set sail for, but never reached, Galápagos.[10]

Having captured a Danish vessel off the coast of Sierra Leone and renaming it *Batchelor's Delight*, a pirate force under the leadership of veteran Pacific captain Edmund Cook rounded the Horn in 1684. Included in the crew were Dampier, Wafer as surgeon, and master and pilot William Ambrosia Cowley. They eventually rendezvoused with John Eaton and the *Nicholas*, which had entered the Pacific in the same year. Dampier (1697: ch. 4) mentions their visit to the Juan Fernández Islands off the Chilean coast, where they had journeyed to inquire about a Mosquito compatriot stranded in 1681 and who had survived the intervening years hunting goats. Dampier describes savannas with great herds that had grown from the three or four goats originally set ashore by Juan Fernández, who had intended to settle there. Without citation, Von Hagen (1945: 187) claimed that the Viceroy of Peru, in order to thwart piracy, had sent men and dogs to Galápagos in order to destroy goats previously introduced by pirates for future provisioning. He was surely mistaken, for Dampier (1697: ch. 6) mentions this claim while anchored near Santa Elena on Ecuador's south coast, but he specifically refers to nearby La Plata Island. Burney (1813–1817: 436) and Gilliss (1855: 41) further confirm that this edict was also delivered for the Juan Fernández Islands, and not Galápagos.

On May 19, 1684, the pirates captured three Panamá-bound supply ships near Trujillo. They fled to Galápagos with their booty, including a wooden image of the Virgin Mary, flour, quince marmalade, "and a stately mule sent to the president." Arriving on May 31, Dampier (1697: ch. 5) mentions staying for 12 days on the second island visited, and setting ashore 5.000 sacks of flour for future provisioning. On the same trip, Cowley (1699: ch. 3) states that "1500 bags of flour with sweetmeats" were stashed on Albany's Bay and other places.

This visit is notable for his chart which first appeared on maps in 1687 (Woram 2016), along with English names he bestowed on the islands and that have remained in varying use since.[11]

Both accounts, plus brief notice by Wafer (1699: 176–177) during the *Batchelor's Delight*'s return visit under the command of Edward Davis to retrieve "500 Packs of Flower, which we had formerly left there upon the Rocks," are important because an isolated Galápagos haven was being used by seafarers for provisioning and careening. Each account details the use of marine turtles, land tortoises, fish, and overly tame birds for food, along with the constant search for potable water sources. Cowley's Albany's Bay is on the northwestern coast of Santiago (Woram 1989: 22), which became a popular careening spot and where later archaeological expeditions found large amounts of colonial pottery.[12] Careening involved beaching the ship after emptying and lowering its topmast. Subsequently turned on its side, the ship's hull could be caulked and scraped of weeds, barnacles, and mollusks (Marx 1992: 199). The fate of Dampier's purloined donkey is unknown, but careening more than likely introduced ship rats into the islands, particularly as their holds were often simultaneously "smoked" in order to kill vermin (e.g., Porter 1822: 237).

The Galápagos Islands continued to furnish safe haven and important provisions for pirates marauding in the Pacific. Smaller-scale raiding by French buccaneers continued after the 1680s, considered the most damaging decade of South Sea piracy (Lane 1998: 154). Repeated visits to Galápagos between 1689 and 1693 were undertaken by a multinational crew of primarily French pirates under the command of Franz Rools, a Zeelander also known as Captain Franco. During this period, they had seized various ships, including the *Begonia* and its stores of flour, biscuit, cheese, and wine, before retreating to their Galápagos hideaway in 1692. Later in 1693, they returned with the *Nuestra Señora del Rosario* and its cargo of wheat, intending to outfit it in Galápagos for the voyage home (Bradley 1989: 168–169). A document describing these adventures by former buccaneer F. Massertie (1600–1701) claims that while careening on the island of St-Bernabé (Santiago) on 18 July 1693, their canoe had returned with three goats. The following day they gathered the goats on the careening island before departing for l'ille Brûlée (Isabela). After again passing by Santiago on 25 July, they eventually unloaded their prize of wheat, tallow, brass, iron, biscuits, flour, meat, mantego cheese, and the goats, and finally departed for the Straits of Magellan in August. This would place the possible introduction of goats into Galápagos well over a century earlier than previously believed.[13] Into the subsequent early half of the eighteenth century, depending upon the changing backdrop of wars and treaties, the islands were regularly visited for careening and provisioning by, including among others, Gouin de Beauchesne in 1700, Rogers in 1709, and Clipperton in 1720, for which we have historical accounts (Burney 1813–1817; Gerhard 1960).

Later Colonial Encounters: Whalers and Sealers

At least by the golden age of Pacific piracy, Galápagos had become a popular venue for careening and provisioning, especially on those few islands affording reliable water sources and accessible approaches. Just as this period was waning,[14] the appearance of new economic forces would dramatically enhance previously established patterns of extraction and introduction, and initiate longer-term human residence on the islands. The eighteenth century saw a new phase of Atlantic whaling, the Southern Whale Fishery, devoted to pursuing highly prized sperm whale oil as the preferred ingredient for a wide range of commodities.[15] After the 1783 Treaty of Paris, largely American and British interests would actively compete for this lucrative trade, which eventually entered into the Pacific (Stackpole 1972).

The *Emilia*, owned by Samuel Enderby and Sons of London, was the first British whale-ship to enter the Pacific early in 1789, and by 1793 as many as 40 English-, American-, and French-based Nantucket ships were operating there (Stackpole 1972: 130). Petitions by London whale merchants compelled HMS *Rattler* to be ordered into the Pacific to reconnoiter and chart Galápagos for commercial ventures. After provisioning on various islands[16] and hunting whales and seals during 1793 and 1794, Capt. James Colnett recommended the islands, particularly Santiago, as convenient for refitting and rendezvous (Colnett 1798).

In short order, Galápagos had become a favorite gathering place for whale-ships and a target for sealers who ranged over the Pacific in search of skins for the Chinese market (Stackpole 1972). Prime whaling grounds were quickly established in the westernmost reaches of the archipelago, particularly in the waters between Isabela and Fernandina where whales seasonally convened in search of squid (Delano 1817: 384; Porter 1822: 138).[17] Various provisioning spots, heavily predicated on their reliability as freshwater sources, were crucial lifelines for whalers who were often at sea for years at a time. Particularly important locations were identified on Floreana, San Cristóbal, Santiago, and Isabela. Whalers regularly visited the islands for acquiring food, gathering wood, ship maintenance, recreation, and overcoming the effects of scurvy. Fish, marine turtles, and birds were procured in great quantities, but few resources were more highly prized than giant land tortoises. Valued for their meat, fat, and a convenient water sack at the base of the neck, they were easy to capture through "turpining"[18] and could survive on board for many months. In a little more than half a century, mariners "loaded their decks with tortoises" and dramatically reduced these once thriving populations, in some cases to extinction (Townsend 1925).

With the onset of hostilities between the United States and England and its North American colonies, Capt. David Porter entered the Pacific early in 1813 aboard the frigate *Essex* in order to disrupt enemy commerce. Armed with a list of whale-ships operating in the Pacific, gratefully provided by American captains liberated by Porter in Peru, he set out for Galápagos as "the most likely place to find them" (Porter 1822: 109). His exploits would prove legendary as he captured a dozen British whale-ships, some carrying arms and wartime letters of marque. He quickly amassed a sizable bounty worth an estimated value of 2.5 million dollars before being defeated in battle at Valparaiso in 1814.

On Floreana, Porter obtained further information of the whaler's itineraries from "Hathaways Postoffice"[19] and provided an account of Irishman Patrick Watkins, possibly the first longer-term resident human in the islands. After leaving an English whale-ship, and prior to fleeing in 1809, Watkins was well known for his locally grown produce which was exchanged with visiting ships for rum or money at "Pat's Landing" on the east side of the island (Porter 1822: 131–135). Watkins was said to have lived in a miserable hut in a valley where he produced "superior quality" "potatoes, pumpkins &c." and chickens on two acres of cultivable ground (Porter 1822: 232). In a later description, surgeon John Coulter (1845: 42) aboard the *Stratford* mentions "four to five acres of excellent land" on which he produced sweet potatoes, pumpkins, Indian corn, melons, hogs and poultry. By the time he escaped the island, Watkins had acquired four "slaves" who lived with him.

Porter (1822: 176) mentions that it was very common for British seamen to abandon their ships, preferring to remain on Floreana. Although he found this extraordinary, he attributes desertion to the onboard tyranny of British vessels. Earlier, sealer Amasa Delano had visited Floreana in 1800 aboard the *Perseverance*. Attempting in vain to land on the three small islands off its eastern shore, Delano eventually made it to a cove on Floreana where he found pegged seal skins and fresh ashes from a fire. After subsequent visits to the island he extolled its virtues for sustaining human habitation, a claim supported by the account of Watkins he had heard years after his earlier visits (Delano 1817: 372). It was not uncommon at this time for visitors to find evidence of previous activity on various islands, especially those with water.[20]

Porter and his crew visited many islands in Galápagos over their approximate six-month sojourn,[21] repeatedly visiting preferred locales in order to provision their burgeoning flotilla of captured vessels. The island of Santiago was a favorite venue, which he considered to have the best harbor in the archipelago (Porter 1822: 208). On August 4, 1831, their flotilla entered Cowan's Bay, the well-known buccaneer haven Porter had renamed in honor of his lieutenant who had died there in a duel, to provision the ships with 14 tons of tortoises. During their stay, four very tame goats and one sheep[22] were left on shore every night, but after several days they had disappeared. Search parties were unsuccessful in locating the animals which had absconded into the highland interior

> where unerring instinct led them to the springs and reservoirs from whence the tortoises obtain their supply; and owing to this circumstance, future navigators may perhaps obtain here an abundant supply of goat's meat, for, unmolested as they will be in the interior of this island, to which they will no doubt confine themselves on account of the water, it is probable their increase will be very rapid. (Porter 1822: 224)

During the decade after Porter's adventures in the Pacific, the Nantucket whalers had assumed control of the Southern Whale Fishery and by the 1840s British involvement had come to an end (Stackpole 1972). Most of the Pacific seal stock had already become seriously depleted by 1830 (Morrell 1832). During the heyday of American whaling between 1815 and its steady decline after 1850, gallons of whale oil and pounds of bone in the millions were harvested by ships from a variety of home ports (Tower 1907: appendix 1). A sudden increase in Pacific whaling, which climaxed in 1846, produced a glut of whale oil on the market, and following financial crisis in 1857 and the outbreak of the American Civil War, the whaling industry never recovered (Tower 1907). Whales were certainly becoming much scarcer, thereby necessitating the increased length and expense of voyages for an already risky business. Crew members were regularly deserting their vessels in California after the 1849 gold rush, cotton had become a more secure investment for entrepreneurs, and petroleum products were becoming cheaper substitutes for whale-based products (Tower 1907: 72). It was during this brief period of whaling's rapid ascendency and decline that Galápagos was annexed by the fledgling Republic of Ecuador.

Republican Period Encounters

Although the Galápagos Islands appeared on some maps as part of the Real Audiencia de Quito, they had never been officially included (Latorre 2012: 37). After the creation of the Gran Colombia in 1822, and during the early decades of Pacific whaling, they were in effect *terra nullius*. Their status was to change shortly after the creation of an independent Republic of Ecuador in 1830. A confederate of Bolívar in the wars of independence, José de Villamil was singularly influential in Ecuador's accession of Galápagos in 1832. Earlier, in 1816, he fought American corsairs attacking Guayaquil and was instrumental in its liberation in 1820. Shortly after independence, Villamil proposed Ecuador's annexation of the islands to President Juan José Flores, and in 1831 formed the Sociedad Colonizadora del Archipiélago de Galápagos (Latorre 2011).

In the following year, Villamil journeyed to the newly christened Archipiélago del Ecuador and its capital Floreana in order to convert the islands into a progressive province symbolically named Asilo de la Paz, with colonies on Santiago, Santa Cruz, and San Cristóbal. He introduced cattle, horses, and possibly sheep[23] into various islands, including San Cristóbal, Santa Cruz, Santiago, Isabela, and Pinta, in order to take advantage of their natural pastures (Latorre 2011: 56). He was significantly assisted in this endeavor by Capt. Nicholas Lawson, a Jamaican who in 1830 had explored the islands, and to whom is attributed the introduction of goats, sheep, and pigs to Floreana and possibly other islands before Villamil's arrival in 1832 (Grant and Estes 2009: 116, 138). In 1833, Coulter visited Asilo de la Paz, which was situated in a valley with a spring close to Floreana's Black Beach. He writes that "this fine valley is of surprising fertility, and produces in abundance sweet potatoes, Indian corn, pumpkins, melons, bananas, plantains, with several other kinds of fruits, and some spices" (Coulter 1845: 53).

Coulter mentions a Swede, Johan Johonson (aka Johnston), who lived on Floreana prior to Villamil's Asilo de la Paz.[24] Previously, Villamil and his "Guayaquil scum" had evicted Johnston and seized his possessions (Coulter 1845: 63). Johnston had been selling fur seal skins and products from his "good supply of poultry and well-stocked vegetable garden" to passing ships for cash and clothing. Earlier, he had also been gifted two donkeys transported from the continent by *Stratford* Capt. Abijah Lock (Coulter 1845: 59–62). Coulter later encountered Johnston and his compatriots continuing their business on Santiago, which they preferred over Floreana as it was more frequently visited (Coulter 1845: 144). Later, in 1835, Lawson had sent 22 men and women from Floreana to Santiago for the purpose of extracting tortoise from the island, where they eventually remained for a number of years (Grant and Estes 2009: 117).

A major factor behind the eventual failure of Asilo de la Paz was the labor force conscripted to fulfill Villamil's dreams. Coulter describes abundant production at the hands of hard-working people, including many women and children. Nevertheless, he portrays "unruly subjects" and "sanguinary black Spaniards" under the governorship of a gentlemanly, yet pompous and self-styled monarch who owned the company store. Prior to his visit, he claims that Villamil had already survived three assassination attempts (Coulter 1845: 53–59). The earliest colonists were composed of mutinous soldiers whose death sentences had been commuted in exchange for exile on the islands, and in later years they were joined by various reprobates from the mainland (Latorre 2012: 38).

In 1835, civil war had erupted in Ecuador, and Lawson replaced Villamil as leader of Asilo de la Paz. During his visit in September of that year, Charles Darwin saw houses irregularly scattered in a flat area where sweet potatoes and bananas were cultivated. He estimated a population of between 200 and 300 inhabitants, "nearly all people of colour, who have been banished for political crimes." The surrounding woods contained many wild pigs and goats but tortoises, which by now had been greatly reduced in number, were relied upon as staple food (Darwin 1845: 375). Eventually resigning his position as Governor in 1837, Villamil was replaced the following year by a tyrannical English soldier, Col. J. Williams. French naval officer Abel Aubert Du Petit Thouars visited Asilo de la Paz in 1838 and details thriving gardens[25] alongside 130 head of cattle, horses, donkeys, several hundred goats, more than a thousand pigs, and "excessive numbers of dogs" (Grant and Estes 2009: 138).

William's reign of terror on Floreana ended badly. General Pedro Mena, another exile from mainland civil conflict who represented Villamil in his absence, departed in disgust for San Cristóbal. Armed uprising by the Asilo de la Paz workforce eventually prompted Williams to flee, and later Villamil relocated the few remaining colonists willing to stay, along with as much livestock as he could, to San Cristóbal; many animals were left behind to turn feral. By 1844, Lawson was residing in Isla Bolivia, a colony established by Villamil on Santa Cruz, where he served as mayor to 21 relocated settlers (Grant and Estes 2009: 116). Floreana continued to be used as a penal colony as 25 convicts resided there in 1845, and by 1848 as many as 90 political exiles were living in 12 derelict houses (Latorre 2011: 79; 2012: 41).

On New Year's Day in 1846, HMS *Herald* and *Pandora* left Valparaiso bound for the Galápagos Islands. The ships had been nominally ordered into the Pacific to improve existing charts of the coastline from Guayaquil to British Columbia. The British admiralty was also interested in the archipelago's suitability as a naval base, instructing Capt. Henry Kellett to survey it for possible colonization (Samson 1999: 74). During their brief visit to the Floreana settlement, which he described as "a few ruined hovels," expedition naturalist Berthold Seeman (1853: 56–59) spoke with William Gurney, an Englishman married to the sister of Sr. Alcé who was apparently in charge. Gurney informed him that many of the exiles had returned to the mainland in the previous year as the Revolución Marcista stripped Flores of his power. Seeman (1853: 58) was impressed by the unbounded fertility of the plantations: "for a mile we walked through enclosures in which Indian corn, melons, bananas, pumpkins, sugarcane, and limes were growing most luxuriantly." No tortoises were found on the island, but wild cattle, pigs, goats, and dogs were abundant. Only two or three of the estimated 2,000 head of cattle on the island were tame; they were easily hunted with dogs and sent to San Cristóbal (Chatham), where a supply station for whalers was being established.

Villamil's early colonizing efforts had included other islands; however, greater attention was now being directed toward San Cristóbal. Coulter (1845: 89–140) had previously explored the highlands of this "thickly wooded land" for two weeks in 1833. After a difficult eight-mile hike on his first day, he encountered open grasslands with "a great many terrapins." Three days later he came across a splendid valley with a coursing stream, luxuriant vegetation, and a "great number of terrapins." Here, he was surprised to find grazing goats which he regularly shot,[26] and later describes finding "immense quantities of coal." As he continued, he found tortoise, turtle, and fish bones scattered around a hut in which three men had lived for more than a year. Four miles inland he encountered recently felled trees in a cleared area and "on further entering this space, there were mustard, pumpkins, melons, Indian corn, sweet potatoes, and tobacco, all growing indiscriminately, and in a very wild state (Coulter 1845: 131)." Coulter then describes a walled clearing overgrown with weeds and a nearby hut "on a comfortable scale" in which he found a man's skeleton covered in skins, which he subsequently buried. If Coulter's claims are valid, then it is possible that goats had been introduced here some time before Porter's introduction of goats onto Santiago in 1813.[27]

Coulter's earlier claim of finding coal on San Cristóbal had attracted the attention of the British admiralty, which was still interested in acquiring the islands, and the HMS *Pandora* was subsequently ordered back in 1847 under the command of Lt. James Wood (Samson 1999: 75). During the previous year both ships had visited San Cristóbal, where Seeman (1853: 61) briefly mentioned "a few poor huts" at Wreck Bay. Here they saw tortoises for the first time and procured wood and small terrapins, 1 by 2 feet, for 6 shillings apiece. Fourteen years after Coulter's remarks about great numbers of terrapins on the island, Wood mentions that settlers took

> as many as 10,000 from this island alone, and so that they are now far from numerous—the wild dogs and the supply of food they afford to Whalers & Settlers will soon destroy all that remain, indeed Mr Gurney [Mayor of Chatham] thought three months work would pretty nearly clear off those that are available for Food or oil. (Grant and Estes 2009: 85)

The San Cristóbal colony tragically comes to light again in 1852 as Ecuador's worst offenders continued to be jailed on Floreana. Renewed conflict on the mainland, coupled with a lack of guards, emboldened an infamous criminal incarcerated on the island to overwhelm the provisioning whaleship *George Howland* in March of that year. Leaving the ship's crew on Floreana, Briones "*el Pirata del Guayas*" and eight others sailed to San Cristóbal where they ransacked the settlement and murdered Mena.[28] Similar problems arose the following year between Floreana prisoners and Gurney, who appears to have remained on San Cristóbal. Villamil continued in subsequent years to pursue interests in reputed guano deposits on the islands, and he formed business partnerships with different North Americans to exploit agriculture and livestock, particularly tallow extraction from feral cattle on San Cristóbal and Isabela.[29] In 1860, he sold a portion of his claims in Floreana to a Frenchman searching for pirate treasure, and he eventually died in 1866 (Latorre 2012: 42–46).

A new commercial venture attracted mainland entrepreneurs to Galápagos with the 1860 discovery of lichens used to produce purple dyes;[30] however, Latorre (2011: 240) cites an 1854 communication to the Interior Ministry that Villamil had already been aware of their existence at least six years earlier. Prosperous Chanduy based entrepreneurs Manuel J. Cobos and José Monroy were able to turn orchilla collection into a profitable venture by growing basic crops and installing provisions on San Cristóbal for their peasant collectors. Nonetheless, by the end of the decade Cobos had run afoul of then president Gabriel García Moreno, who subsequently awarded exclusive orchilla collection rights to Santa Elena–based businessman José Valdizán. After making arrangements with the Villamil family, Valdizán reestablished agricultural operations on Floreana, yet again with convict labor. Tragedy struck once more, as Valdizán was assassinated by members of his workforce in 1878. The island was quickly abandoned, with many laborers relocating to San Cristóbal (Latorre 2002, 2012: 49–59).

Naturalist Teodoro Wolf spent a total of six months in Galápagos during two trips in 1875 and 1878 on Valdizán's ship *Venice* when he visited various islands including Floreana. The major purpose of Wolf's second visit, after his appointment as State Geologist, was to find guano deposits on the islands (Sevilla 2017b). Also, his summary account reports on their potential for colonization, particularly as a naval station at Post Office Bay, with the anticipated opening of a Panama Canal (Wolf 1887). Both times, Wolf visited Asilo de la Paz, finally departing the islands only three days before Valdizán's assassination in July of 1878. He mentions a flourishing settlement guarded by living fences and lime thickets to keep out wild cattle. He was surprised by the island's great fertility, and how temperate crops had acclimated in the tropical gardens used to provision the camps of orchilla collectors.[31]

Wolf (1887: 20–22) offers an account of introduced animals, which had become perfectly acclimated on the islands since Darwin's brief 1835 visit. Land tortoises were by now severely reduced in number and completely extirpated on islands like Floreana. He mentions great wild herds of large cattle estimated at 800 to 900 head on Floreana, 2,000 to 3,000 on San Cristóbal, and more recent introductions into the southern mountains of

Isabela. Wild cattle were hunted on San Cristóbal for meat and hides consumed locally and to provision whalers. Settlers were also commencing to domesticate cattle by allowing their corralled cows to interbreed with wild bulls in highland pastures. Wild horses on Floreana had been tamed by Valdizán; donkeys, easily tamed by orchilla collectors, were abundant on Floreana, San Cristóbal, Santa Cruz, Santiago, and Isabela; herds of goats were found on Floreana and San Cristóbal, with isolated introductions on Santa Fé and Española;[32] wild pigs were on all the large islands, especially Santiago; chickens were found only in the hills and forests of Floreana; cats lived in coastal lava tubes, feeding on marine life, tame birds, and introduced rats and mice; and the numerous packs of dogs were abundant and easily tamed. Subsequent visits to Floreana, after the island was abandoned, provide descriptions of a thriving feral landscape brimming with potential for subsequent human colonization (Latorre 2012: 58–59).

Cobos, who had earlier departed for Mexico's Lower California in pursuit of commercial ventures centered on orchilla collection, renewed his focus on San Cristóbal around the time of Valdizán's death. The trajectory of his business interests had changed after the 1875 assassination of García Moreno, coupled with a waning demand for orchilla[33] and involvement in some legal entanglements in Mexico. From its previous modest beginnings on a small croft once occupied by a negro and populated with laborers from the mainland (Mann 1909: 28), along with later refugees from Asilo de la Paz, the Hacienda El Progreso would rapidly become the apogee of nineteenth-century development on the islands. It would continue into the subsequent century, and after Cobos's assassination in 1904, as a hallmark of human habitation on the Galápagos Islands. A detailed history of El Progreso is discussed in the following chapter.

Hacienda El Progreso was soon joined by another colonizing effort under Antonio Gil, a Guayaquil politician and friend of fellow costeño President Eloy Alfaro. Attempting to replicate the success of El Progreso, Gil originally set out to resuscitate Asilo de la Paz in 1897, but he later directed his energies toward Isabela in 1900 (Latorre 2011: 22). Gil established the coastal town of Puerto Villamil and the nearby highland settlement of Santo Tomás de Berlanga in order to grow agricultural products and export hides of feral cattle, mined lime and volcanic sulfur, and rendered turtle oil. Two early visits, one by Nicolás Martínez who accompanied Antonio Gil's son aboard the *Tomasita* in 1906, and the other by Alexander Mann on his motor-schooner *Scotia* in 1907, describe the

Isabela colony, which already had 200 inhabitants by 1905. In both narratives, Martínez and Mann leave San Cristóbal and provide brief mention of passing islands. Despite its description as a desolate and waterless rock, Santa Fé sustained numerous goats allegedly introduced by Cobos. Evidence for human habitation was mentioned on Santa Cruz, which was very important for its cultivable land and the many burros and goats that lived there.[34] Floreana, seen off in the distance, was reported to have only plantation ruins and domestic, presumably feral, animals (Mann 1909: 41–43; Martínez 1915: 71–73).

Puerto Villamil was described as a depot for export products including hides, sulfur, and tortoise oil. Some 50 or more inhabitants were employed in loading and unloading ships, fishing, burning shells for calcium, and rendering oil. Gil had constructed a tannery using local forest tannins to process cattle, sea lion, and dog hides. Both visitors ventured some 15 km inland via a three hour mule ride to Santo Tomás. Mann describes the difficulty of passing over sharp volcanic outcroppings that had to be covered in bagasse, or residual sugarcane pulp. Gil had constructed a small, oxen-powered mill for the production of locally consumed syrup and alcohol, most of which was imported from Hacienda El Progreso. Both remark on the verdant fertility of the land with products superior to those of San Cristóbal.[35] The settlement, dominated by the hacienda house and its vineyards, was populated by 150 men, women, and children, principally employed in mining, gardening, cattle tending and hunting, cheese making, breeding and fattening hogs, and tortoise oil extraction (Mann 1909: 44–46; Martínez 1915: 81–92).

Both men were clearly impressed by the extensive herds of domestic and wild cattle grazing in grass-covered highland pastures that started within 3 km of the coast. Village inhabitants were engaged in domesticating wild cattle, found particularly in higher-elevation pastures, by penning those that had returned with domestic decoys to their corrals. Mann (1909: 52) estimated 8,000 to 10,000 head of "very fine animals, large, massive, and clean-skinned" cattle, whereas Martínez (1915: 86) projected as many as 20,000 wild and 2,000 domestic varieties, exclaiming that the pastures around Sierra Negra Volcano were sprinkled by thousands of points of color of wild cattle. Both visitors were alarmed by the "terrible plague" of wild dogs that swarmed the pastures in crowds of hundreds.[36] Gil had also introduced "fine races of sheep" to the island, and at some distance below the cattle pastures were piggeries, and poultry farms with hundred of chickens and ducks (Mann 1909: 52–56; Martínez 1915: 86–94).

Early Twentieth-Century Encounters

Aided by naval support from the United States, Panamá seceded from Colombia in November 1903 and, within weeks after declaring independence, signed the Hay-Bunau-Varilla Treaty, creating the Panama Canal Zone. The treaty granted land sovereignty and exclusive canal construction rights to the United States, which in 1904 purchased existing assets of the French New Panama Canal Company (Ameringer 1963). Ten years later, the completed canal linked the Atlantic to the Pacific,[37] significantly facilitating European accessibility to Galápagos. In subsequent years the islands would be increasingly visited by an assortment of scientists, adventurers, colonists, and disillusioned émigrés.[38]

In the decades between the two world wars, the Galápagos Islands were regularly visited by a steady stream of research expeditions from the United States and Europe, often funded by wealthy benefactors. Most were involved in some kind of biological collection, which was significantly prompted by the scientific community's increasing alarm at the rapid disappearance of iconic land tortoises. The canal also opened access to the private yachts of adventure seekers who frequented the islands and occasionally supplied accounts of their visits.[39] Serious new attempts to colonize the islands began in 1925 in the form of three separate, yet related Norwegian proposals, although interest in settling Galápagos appeared as early as 1884.[40]

August Christensen, son of a Norwegian whaling pioneer and eventual consul in Guayaquil, enthusiastically promoted Galápagos in popular publications around the time of the canal's completion. With whaling rights obtained for the Ecuadorian and Peruvian coasts, he organized La Compañía de Floreana, a joint stock company, which was to use the island as a whaling station. A series of enthusiastic accounts by Norwegian journalists traveling in 1922 as guests of Manuel A. Cobos describe their brief stay on Floreana with its teeming population of wild pigs, goats, cattle, and abundant orange and lime thickets. The island was promoted for a variety of activities, and in 1925, the schooner *Floreana* arrived with more than 300 tons of supplies to construct a prefabricated Casa Matriz, or main building, dock, steel tracks, and water systems.[41] The fledgling settlement was divided into two teams, one dedicated to commission-based seafaring, and the other to cattle ranching and fishing, with future plans for coffee, tobacco, and banana plantations, whaling, canning, and a maritime supply depot. Surviving into the New Year, and after receiving Christmas gifts of oxen, cows, horses, donkeys, and pigs from Cobos, the colony finally broke up in January of 1927 (Lundh 1999, 2006a; Hoff 1985).

An early enthusiast of the failed Floreana venture, Olaf Eilersten organized La Colonía de Santa Cruz, arriving August 1926 in Academy Bay with 45 colonists aboard the *Ulva*. They were not alone on the island, as some 7 km further inland laborers were attempting to establish a cattle ranch on previously abandoned farms attributed to Cobos's orchilla collectors. Shortly after the colonists' arrival, the imported cattle from San Cristóbal were abandoned, after which they headed for highland pastures, where they eventually bred with cows released in 1927 by Norwegian farmer Jacob Horneman.[42] The ambitious coastal colonists erected seven prefabricated wooden houses, dynamited a channel into a back-bay lagoon, and constructed a pier with steel rails leading to a canning factory for processing mullet, sea turtles, and lobster. A second vessel with colonists never made it further than Panamá; most of its passengers returned to Norway or dissipated into tropical America, but four found their way to Galápagos. Disagreements and lack of capital led to failure, and by 1932, only two colonists remained on the island (Lundh 1999, 2006a; Hoff 1985).

A third Norwegian colonization with 83 women, men, and children left Oslo in September 1926 to settle Campo Noruego, where 14 prefabricated highland houses were erected on San Cristóbal, primarily for agricultural pursuits.[43] Through negotiations with Cobos and Rogerio Alvarado, each settler received 20 hectares of land, free rent for two years, commissary credit, and cheap access to labor and draught animals. Returning to Guayaquil, project organizer Harry Randall discovered that neither Cobos nor Alvarado held any legal claim to lands or cattle. Nevertheless, the scourge of red ants and pigs devastated their temperate crops, and by 1930 few of the Norwegian settlers remained; a few relocated to Santa Cruz.[44]

Although a number of planned settlement projects never succeeded in even reaching Galápagos, the few Norwegian settlers who remained on Santa Cruz would be joined over the years by others from Europe and the Ecuadorian mainland. Among the early newcomers were the celebrated Angermeyer brothers, who left Germany to colonize the island in 1935. Their early survival required relocation to the highlands, which they accomplished by transporting their possessions with captured donkeys, eventually finding an area with a beautiful view and wild avocados, oranges, and bananas. They hiked for many miles to areas where pirates had planted bananas, and after spending the night loaded their donkeys with wild lemons, avocados, papayas, oranges, and green banana stems. They lived off wild boar, tamed a cow for milking, obtained some chickens, and stored their provisions in rat-proof drums (Angermeyer 1989: 125, 153).[45]

Floreana apparently had one permanent resident, Hugo, in 1929 when disillusioned German couple Friedrich Ritter and Dora Strauch arrived with plans to establish a new life on the island. Hugo exploited the highland's "herds of roaming wild cattle" and wild boar which he was "in the habit of killing to his heart's content" (Strauch 1936: 39, 42). He had been stationed there to process and collect cattle hides by Paul Bruun, who was involved in the earlier Norwegian colonization, and who would return in the following year to briefly reoccupy the Casa Matriz in yet another failed fishing venture. Ritter and Strauch had been put ashore with an assortment of supplies deemed necessary for survival, including "seeds of all vegetables and fruits which might be made to grow in that climate" and a crate of chickens (Ritter 2015: 12), the latter apparently unnecessary as feral chickens had already begun to nibble their imported plant shoots.

Immediately setting out for the highlands they encountered a clear spring in the center of a volcanic crater, where "all about grew tropical fruits in abundance, including bananas, oranges, pineapples, lemons, guavas, papayas and many others which I could not identify" (Ritter 2015: 16). In their new home, christened Friedo, Garden of Peace, "we had for neighbors a vagrant colony of wild cattle, wild hogs, wild asses, wild dogs, and wild tabby cats" that roamed the entire island (Ritter 2015: 21). Feral animals, which continued to use the paths created by a once flourishing population of wild tortoises, now extirpated by dogs, would prove challenging as they constantly had to battle cattle, asses, hogs, dogs, and cats to protect their agricultural efforts.[46]

The growing notoriety of Friedo[47] attracted other Europeans, including the Wittmer family in 1932, and later that year the infamous "Baroness" Wagner de Bosquet and her male retinue, who arrived with cows, donkeys, flocks of ducks, hens, and turkeys, rabbits, pigeons, and 70 hundredweight of cement to build a planned Hacienda El Paraiso (Strauch 1936: 159).[48] The Wittmers settled near the site of the abandoned Asilo de la Paz and suffered similar depredations from cattle, pigs, donkeys, dogs, cats, and rats, especially in the early years of garden establishment.[49] The Wittmers quickly had to learn how to preserve the excessive quantity of meat they obtained from wild boar and bull. Nights were filled with animal sounds, and days with their appearance: "a big herd of cattle and their calves moved peacefully over the pampa, pigs and donkeys mingling with them in easy familiarity. It was an idyllic scene, perfect for the first Christmas on 'our island'" (Wittmer 1961: 52).

The Baroness disappeared mysteriously in March 1934, and the last of her entourage left Floreana in July. Ritter died of food poisoning in November of the same year, after which Strauch returned to Germany. The Wittmers, whose descendants would comprise the longest-residing family in island history, were soon joined by an American couple. Ainslee and Frances Conway were urged in 1937 to first settle on Santiago as the best of the habitable islands. They did not consider San Cristóbal. Santa Cruz had little meat but plenty of rain, good land, and jungle in its highlands. Floreana had grasslands, plenty of pig and cattle, but little rain. Isabela was fine but had prisoners and too many people who thought they owned everything. Santiago was preferred; it had no people and had not been planted, but it had goats, pigs, burros, horses, and cattle, suggesting that it had good springs (Conway and Conway 1947: 34).[50]

By this time, Santiago had already been an important human destination for more than 250 years. Darío Egas Sánchez, who had owned the west side of the island since 1922, began salt mining operations in 1926 under contract from the government after salt pans in Santa Elena had flooded (Lundh 2006a).[51] Salt continued to be extracted in 1931 by Paul Bruun, who had returned to Floreana to establish the necessary infrastructure for shipping dried fish to the continent. He enlisted the aid of Temple Utely, who visited Santiago that year on his yacht *Inyala*, describing a salt extraction operation with tracks for an oil-powered engine and 70 laborers who had mutinied for lack of water. Within a month, Bruun perished in a boating accident on the breakers off Isabela, and his planned business venture dissolved (Utely 1938).

The Conways were eventually forced to leave Santiago as the authorities could not guarantee their protection on an island deemed unfit for humans. Upon relocating to Floreana, the Wittmers recommended they live deep in a lemon jungle and close to water on the old site of Asilo de la Paz. On the way to their new homestead they described the Floreana landscape, visiting the abandoned site of Friedo with its scattering of fruit trees and palms, and the "faerieland" of the Wittmers with its "typical German vegetable garden" and caves converted into a smoke house and pig pens.[52] While clearing land and waiting for their "initial emergency garden" of fast-growing crops to mature, the Conways lived on products of the past, especially lemon, orange, guava, and avocado jungles, feral pigs hunted by dogs, and cows, which were preferred over bulls. "Meat kept coming into our camp faster than we could dispose of it" (Conway and Conway 1947: 165). Later plantings of subsistence staples were followed by a host of flowers for decoration, producing gardens with more than 60 varieties of plants from all over the world (Conway and Conway 1947: 266).[53]

By the onset of the Second World War, the Galápagos archipelago had been regularly frequented for centuries, during which at least five major islands experienced some significant historical entanglement with humans. Since their establishment in the previous century, enduring settlements on San Cristóbal and Isabela continued to pursue agriculture and resource extraction. Floreana and Santiago served vital roles for seafarers over the course of centuries, in addition to serving as insular locations for sporadic settlement. Although regularly visited somewhat later than other islands, the highlands and southern coast of Santa Cruz also became destinations for human settlement. As hostilities erupted in the Pacific theater, Baltra, a smaller island lying directly adjacent to the northern coast of Santa Cruz, would become the next site of intensive interest.

Since its annexation by Ecuador in 1832, Galápagos was coveted by different countries, particularly the United States, for various reasons, including payment of war debt, loan collateral, guano and coal concessions, and as a strategic naval and supply depot. Shortly after the Panama Canal's completion, the United States proposed a 99-year, $15 million lease that was rejected by the Ecuadorian public because of US involvement in Panamanian secession. The United States continued its interests in pursuing the archipelago's strategic location, culminating in the construction of Base Beta on Baltra in December 1941.[54] It included a deepwater dock, fuel depot, roads, two asphalt air strips (one more than 1.8 km in length), an air base, hangars, 200 buildings with barracks for 1000 soldiers, offices, churches, cinema, hospital, bowling alley, outdoor beer garden, and the world's largest bar.[55] Radar outposts were also placed on Española, Isabela, and nearby Daphne islands (Epler 2013; Idrovo 2008). Despite US efforts to extend the lease, its forces departed on July 15, 1946, and island dwellers were invited to scavenge the abandoned material. In a February 9, 1944, Memorandum for the Secretary of State, President Roosevelt suggested that an Ecuadorian proposal be submitted to the Pan American Republics to protect the islands in return for rent, and that under Ecuadorian sovereignty both countries would police the islands for protection of the South American continent and the Panama Canal:

> These Islands represent the oldest form of animal life and should, therefore, be preserved for all time as a kind of international park, the title of which would remain in the Ecuadoran Government but the operation of which would be (1) in the scientific area run by a committee representing each of the twenty-one Republics and (2) the policing area which would be run jointly by the United States and Ecuador. I think it would be safe to ask Ecuador to abandon any thought of agriculture or cattle raising in the Islands. This would not amount to anything much anyway. (Woram 2016)[56]

Current Human Geography in Galápagos

In 1950, four islands in Galápagos housed a resident human population estimated at 1,346, and by 1962 this number had almost doubled (Neira 2016). Rapid and sustained growth, however, began in the mid-1970s as immigrants were attracted to the opportunities afforded by a dramatic increase in tourism, coupled with economic and political turmoil on the mainland.[57] Human population in Galápagos more than tripled between 1974 and 1998, from 4,078 to 15,311, due largely to immigration (Epler 2007).

In 1972, more than a decade after its declaration as a national park, Galápagos officially joined Ecuador as a province, which according to the 2010 census boasted a population of 25,124.[58] Today, approximately 96.7% of the islands' total land area of 799.54 ha is set aside as a park, the remaining 3.3% regarded as Zones of Special Use (Figure 2). These coincide with the historically colonized areas of San Cristóbal, Santa Cruz, Baltra, Floreana, and Isabela; however, 85% of the resident population is today concentrated in the major coastal town of each inhabited island (INGALA 2005).

The province is divided politically into three cantons. Cantón San Cristóbal includes the eastern islands and Floreana, with 7,475 residents, more

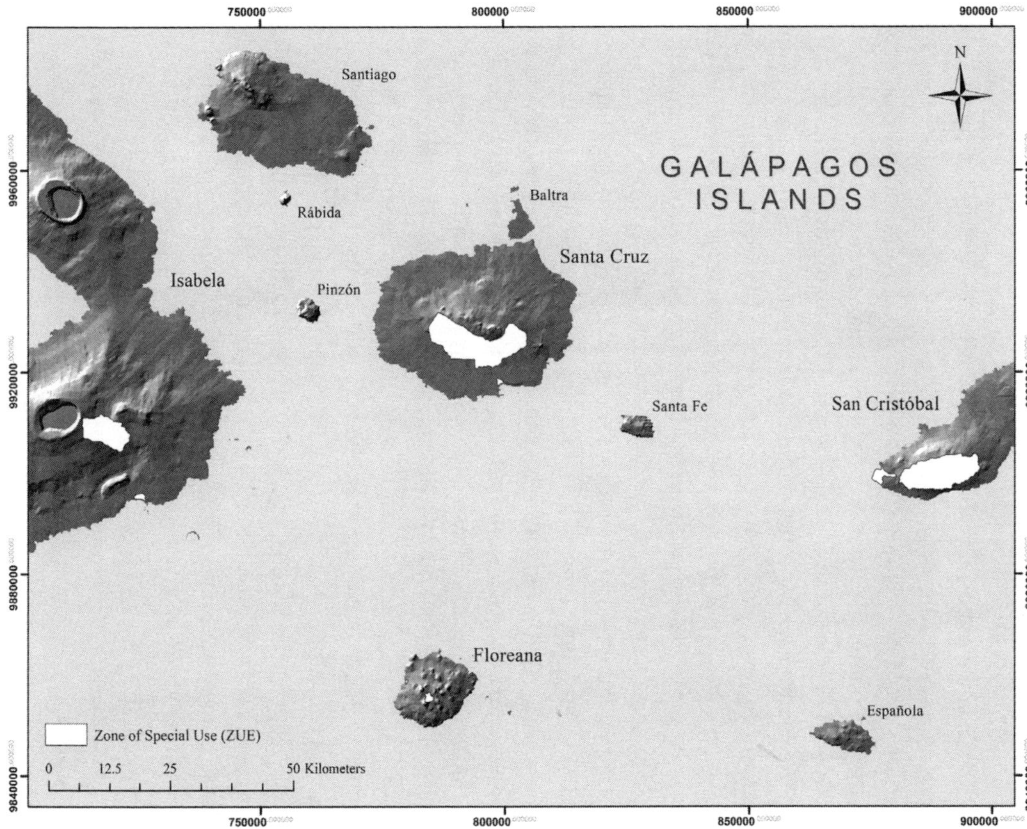

Figure 2. Zones of Special Use (ZUEs) in the four populated islands of Galápagos National Park. Map Datum: UTM 15S WGS 1972, created by Rylee Harlos.

than 75% residing in Puerto Baquerizo Moreno (5,539), the provincial capital on San Cristóbal's Wreck Bay. Throughout the latter half of the twentieth century, San Cristóbal housed the largest continuous resident population on the islands, and Puerto Baquerizo Moreno, serviced by its US military water pipeline, was the archipelago's largest town until the 1990s. Today, it is the site of the island's second airport, constructed in 1986, a naval base, government offices, hospital, university campus, hotels, restaurants, shops, and an economy involved principally in tourism, administration, and fishing.

The balance of San Cristóbal's population resides in the humid highlands, where farming and cattle ranching for island and tourist consumption predominate. Many coastal dwellers also own highland property in numerous scattered farmsteads, in a few small hamlets, or in and around the town of El Progreso (497), which lies approximately 7 km inland from Puerto Baquerizo Moreno. Floreana has been continuously occupied since the Second World War by a few resident families. Today, the island has approximately 30 houses, mainly in Velasco Ibarra (109) on Floreana's Playa Negra or Black Beach, and a third of its population is located in the highlands, where farming continues to be pursued.[59]

Cantón Santa Cruz includes the central and northern islands with 15,393 inhabitants, almost 60% residing in the cantonial capital Puerto Ayora (9,208) on Academy Bay. The balance of the population occupies scattered farmsteads and the smaller highland towns of Santa Rosa (461, including Baltra) and Bellavista (1,881), approximately 7 km from the coast. After the Second World War, most residents of Santa Cruz subsisted on fishing, farming, cattle ranching, and trade with the American tuna fleet that fished in the area until the mid-1970s (Epler 2013: 44). After the tourism explosion, Puerto Ayora became the largest town and major tourist hub in the islands, and today it sports administrative buildings, a hospital, banks, a post office, hotels, restaurants, shops, and the Charles Darwin Research Station. Most visitors to Galápagos arrive at the airport on Baltra, which received its first commercial flight in 1963, and are then bused southward to Puerto Ayora via ferry over the Canal de Itabaca and a paved highway that transects Santa Cruz. Baltra also houses military personnel and serves as a fueling station for island-based boats.

Cantón Isabela includes the western islands with 2,256 inhabitants; most (84%) reside in the town of Puerto Villamil (1,892) nestled inside Bahía Villamil on the southern tip of Isabela. Approximately 140 people live in the highland town of Tomás de Berlanga, and the remainder in scattered farmsteads. A notorious penal colony was established on Isabela in 1944, and by 1950 the majority of the island's 500 inhabitants were prisoners incarcerated at Puerto Villamil, Santo Tomás, Alemania, six hours inland, and Porvenir, the site of an abandoned radio station (Epler 2013: 49). The penal colonies were abolished in 1959,[60] and today island inhabitants are engaged in fishing, farming, and tourism. Puerto Villamil has an intra-island airport, constructed in 1996.

Historic Introductions, Naturalized Aliens, and Landscape Transformation

Over recent centuries, Galápagos ecosystems have been intentionally and accidentally altered by human activity through the coupled depletion of native organisms and supplementation with alien introductions. Many introduced alien organisms exhibited transient residency and others became naturalized, with some implicated in landscape transformation. An oft-cited statistic proposes that Galápagos still retains 95% of its original pre-human biodiversity (Watkins and Cruz 2007); however, a growing list of alien plant and animal introductions, estimated at 1,445 a decade ago, continues as the greatest external threat to the park's terrestrial biodiversity (Atkinson et al. 2012).[61] Although the rate of alien introductions, especially of plants, has been characterized as exponential and strongly coupled with recent human population explosion (Mauchamp 1997), rates in Galápagos appear similar to trends found in most islands, and more likely suggest multistage linear increase (Tye 2006).

Transformed landscapes, particularly in the highland interiors of islands endowed with the requisite resources for supporting certain naturalized populations, are among today's conservation priorities. Modification through human activities is particularly severe in the humid and very humid vegetation zones of San Cristóbal, Santa Cruz, Floreana, and southern Isabela (Watson et al. 2010). Having evolved in their absence, the insular biotas of Galápagos are particularly susceptible to larger mammalian herbivores,[62] which have had centuries to establish extensive, free-roaming, feral populations in these areas (Coblentz 1978). Whether in the presence or absence of humans, a number of introduced mammals have become key alien transformers driving ecological change, most dramatically in the humid interior highlands of historically inhabited islands.

Naturalized alien organisms can transform native landscapes, both through direct predation on animals and plants, and by indirect resource competition and habitat alteration, which can simultaneously degrade conditions for certain organisms while enhancing opportunities for others. Feral pigs, which are found on four inhabited islands,[63] prey directly on vertebrates and invertebrates. Extreme generalists, pigs eat native reptiles and birds and are particularly damaging to nesting species by rooting up nests and eating hatchlings and eggs.[64] Introduced dogs and cats were found on nine islands and are naturalized on San Cristóbal, Floreana, Santa Cruz, and Isabela, where they are serious threats. Both are implicated in the decline of native invertebrates, reptiles, birds, and mammals, through hunting and nest predation.[65] Although impossible to substantiate, exotic rats may have been the earliest introduced alien on the islands and are today particularly widespread, having invaded at least 36 islands, where they prey on the eggs and young of reptiles and birds (Phillips et al. 2012: 470).[66]

Naturalized populations of large, free-roaming, feral grazers and browsers can compete with endemic animals by simultaneously consuming native plants and trampling their nests, eggs, and young. Goats are particularly damaging, for as extreme generalists, "very little palatable forage escapes them" (Coblentz 1978: 283). Goats often preferentially browse native over introduced plants and are major threats to a significant portion of plants currently in peril.[67] The ecological legacy of goats is particularly widespread and severe throughout the Galápagos archipelago, as they were introduced early and often on many islands, and they are markedly adaptable and fecund.[68] Goats, pigs, donkeys, and cattle have seriously damaged *Scalesia* and *Miconia* vegetation on inhabited islands (Hamann 1984; Kastdalen 1982; Schofield 1989). Donkeys and cattle regularly trample the nesting sites of tortoises, iguanas, and terrestrial-nesting birds like Dark-Rumped Petrels (Carrion et al. 2007; Phillips et al. 2012).

Cattle are particularly capable of transforming island habitats by changing the composition, structure, and dynamics of communities. Grazing can inhibit species recruitment, alter floristic diversity, and control the timing and nature of ecological succession. Repeated grazing creates "grazing lawns" in which plants are maintained in a juvenile and rapid growing state (Cuevas and Le Quesne 2006; Hobbs 1996). Grazing, browsing, and trampling by cattle, goats, and donkeys can alter understory structure and open up closed vegetation, further enhancing invasion by opportunistic organisms, especially annuals and grasses. Heavy grazing often favors invasive plants, the majority of which are herbaceous weeds whose seeding and dispersal mechanisms are better adapted to novel conditions (Hamann 1984; Trueman et al. 2013).[69]

The transformative nature of certain naturalized aliens is further demonstrated by landscape change after their removal, resulting in the regeneration of previous conditions and/or increased novelty.[70] The effects can become dramatically pronounced in areas where eradication programs of key transformers have taken place. Outcomes can include the renewal of previous vegetation through the ability of seedlings to rejuvenate and mature in the absence of grazing pressure, which restores habitat for native animals.[71] Removal of an alien transformer can also lead to cascading effects, often resulting in the increased opportunity for, or establishment of, another alien invasive.[72] Recent estimates suggest that the majority of Galápagos floras are alien, pronouncedly in the humid highlands of inhabited islands where some have become transformative. Many succulents (e.g., agave, airplant), flowering and garden plants (e.g., daisy, datura, cucumber), vines (e.g., passion fruit), shrubs (e.g., blackberry, nettle, sage), trees (e.g., guava, laurel, avocado, cedar, balsa, cinchona, citrus), and grasses (e.g., couch, bunch, elephant) have invaded large areas and adversely affect natural ecosystems (Tye at al. 2002).[73]

Eradication efforts have been undertaken and are currently under way or planned on various islands.[74] Some have met with spectacular success and others have failed. Eradicating different organisms can require different approaches in different contexts; lack of success can generally be attributed to operational failure (e.g., remnant breeding populations survived, distributions were poorly known), species biology (e.g., some species are difficult to eliminate); lack of resources or insufficient time (successful eradication requires money, adequate tools, labor, and time), and sociopolitical support (e.g., lack of will, permission, and/or conservation ethic) (Atkinson et al. 2012; Gardener et al. 2010). Today, the heaviest concentrations of alien invasive organisms are found in the agriculturally productive interior uplands of inhabited islands, and particularly in San Cristóbal's humid and very humid highlands (Watson et al. 2010). These areas supported the most sustained and intensive human productive effort in Galápagos' history. An unwritten but often repeated sentiment is that conservation priorities should perhaps best be focused elsewhere as restoration efforts on the island's southern end are hopeless.

3

Manuel J. Cobos, San Cristóbal, and the Hacienda El Progreso

The legacy of earlier human occupation in Galápagos continues today in the extensively transformed landscape of San Cristóbal. The island's accessible and sheltered bays, verdant interior highlands, and reliable freshwater sources contributed to its popularity, first as a provisioning destination for seafarers, and eventually as safe haven for human colonization. At least by 1830, domestic animals were introduced onto San Cristóbal, after which small human colonies began to take hold. These would be regularly supplemented by refugee colonists fleeing the periodic turmoil on Floreana, and during the 1860s commercial interests prompted further settlement and provisioning for orchilla collection. After another assassination on Floreana in 1878, Manuel Cobos established his personal presence on the island to create the apogee of early commercial production in Galápagos.

San Cristóbal

San Cristóbal is the fifth-largest and easternmost island in the Galápagos archipelago. Located 960 km to the west of the Ecuadorian mainland, this 558 km^2 island is one of the oldest in a chain of emerged volcanoes produced through seafloor spreading. Each island was formed serially from east to west by underwater eruption as the crust and mantle of the southeasterly rafting Nazca plate melted while passing over a stationary and intense hot spot (Jackson 1993). Geological dating suggests that the initial emergence of San Cristóbal may have begun almost 2.5 million years ago.[1]

The island was formed by two petrochemically similar volcanoes during temporally overlapping periods. An older shield volcano dominates the southwestern portion of San Cristóbal and a younger fissure-fed volcanic center, demarcated by an intermediate boundary, dominates its northeastern half (Geist et al. 1986).

The interior highlands of the island's southern portion attain a maximum elevation of nearly 900 m asl on Cerro San Joaquin, a parasitic cone lying to the northwest of Laguna El Junco, the water-filled explosion crater at the summit of an ancient caldera rim. Together with nearby ponds and a regular stream coursing down the volcano's southern flank that enters the ocean at Freshwater Bay, El Junco forms an important and rare freshwater habitat in Galápagos (Colinvaux and Schofield 1976). The moist uplands receive much greater amounts of precipitation during two distinct annual seasons modulated by surrounding oceanic currents, together with trade winds governed by the intra-annual north-south migration of the Intertropical Convergence Zone (ITCZ) of low atmospheric pressure lying to the north of Galápagos.

During the hot season between January and May, when the ITCZ is nearer, trade winds diminish, facilitating the arrival of warmer northern ocean currents. Ocean and air temperatures are higher, and most days are sunny. Precipitation is orographic, as mild east-southeast winds bring higher rainfall with increased elevation in the form of variable small and short convective storms. Up to 660 mm of rain can fall in the lowlands, compared with 1,263 mm in the highlands. During the cool season between June and December, when the ITCZ moves northward, colder southeasterly trade winds blow consistently over southern cold oceanic currents circulating offshore. Ocean and air temperatures are lower, and most days are characterized by persistent and extensive stratus clouds as the cooled air is trapped underneath warmer air. Precipitation in the form of rainfall or condensed *garúa* fog occurs almost every day where land surfaces rise with increased elevation through this inversion layer. Up to 10 mm of rain can fall in the lowlands, compared with 67 mm in the highlands. This annual pattern is regularly punctuated by the Pacific-wide El Niño Southern Oscillation, when El Niño and La Niña conditions bring protracted hot and cool season conditions respectively (Trueman and d'Ozouville 2010).

The entire range of altitudinal vegetation zones, which have been used to characterize the islands, is particularly comprehensive where there is sufficient elevation to support humid highlands. However, these can be difficult to corroborate on San Cristóbal, owing to extensive human landscape modification on the southern end of the island. Traveling from the coast into the interior highlands along San Cristóbal's main road, elevation begins to rise more rapidly than on southern portions of Santa Cruz where these intergrading vegetation zones were described (Wiggins and Porter 1971).[2] The vegetation of a narrow Littoral Zone is directly influenced by salt and dominated by shrubs or small trees and salt-tolerant herbs and grasses. An Arid Zone dominated by deciduous plants, Prickly Pear (*Opuntia* spp.), Candelabra (*Jasminocereus thouarsii*), and Lava (*Brachycereus nesioticus*) cacti, shrubs, herbs, and smaller trees, including Palo Santo (*Bursera graveolens*), lies immediately inland to an elevation of 80 to 100 m asl. Forests of palo santo end on San Cristóbal at around 55 m asl, where Woody Croton (*Croton scouleri*) increases along with trees of Manzanillo (*Hippomane mancinella*), Matazarno (*Piscidia carthagenensis*), and Pega Pega (*Pisonia floribunda*).

An extensive and greener Transition Zone incorporates xerophytic plants of the Arid Zone with moister mesophytic plants of the higher Scalesia Zone. A few ferns survive and grasses become more abundant, with closely spaced, taller trees, including pure stands of manzanillo accommodating lichens and bryophytes. The lower edge of a Scalesia Zone begins between 180 and 200 m asl and is characterized by small trees of *lechoso* or Tree Scalesia (*Scalesia pedunculata*), *Guayabillo* or Galápagos Guava (*Psidium galapageium*) mixed with pega pega, Cat's Claw (*Zanthoxylum fagara*), and a prominent shrub understory with herbs, vines, and epiphytic ferns, orchids, liverworts, and mosses. Scalesia forests are engulfed by *garúa* during the cool season and assume the appearance of an island rain forest.

A Brown or Zanthoxylum Zone has been suggested as transitional between the Scalesia Zone where it grades into the higher Miconia Zone. Here, epiphytes that can turn brown in drier seasons cover dominant Palito Negro (*Tournefortia rufo-sericea*) and Romerillo (*Macraea laricifolia*) shrubs. Elevations above 500 m asl form a Miconia Zone dominated by open stands of Cacaotillo (*Miconia robinsoniana*), along with clubmosses, tree ferns (*Cyathea weatherbyana*), and Chontillo (*Pteridium aquilinum*) as rainfall diminishes and soils are slightly more acidic. Also called the Pampa or Grassy Zone, an upper Fern-Sedge Zone can persist to the tops of island peaks with narrow-leaved sedges, clubmosses, ferns, lichens on exposed surfaces, and occasional tree ferns in swales and pockets. Soils are deep, slightly acidic, and relatively high in moisture retention, and peat bogs can persist in protected areas.

Vegetation cover and rapid weathering of volcanic deposits in the perpetually damp highlands have formed soils of sufficient depth for agriculture, commencing with the gray-brown podzolic and brown forest soils of the Scalesia Zone (Geist et al. 1986; Stoops 2014: 11; Wiggins and Porter 1971: 16). Park boundaries in the southern end of the island roughly coincide with the lower limits of this zone at around 200 m asl. The boundary is a clearly visible line separating grayish mauves of deciduous vegetation at lower elevations from the verdant greens of forest canopy and open grasslands above. The entire highland area supports agricultural activity, where the principal mechanisms driving ecological structure are anthropogenic.

A modern paved road, recently improved in 2014, bisects the interior Zone of Special Use on the southern end of the island (Figure 3). Beginning directly east of Puerto Baquerizo Moreno, it climbs into the highlands, skirts the southern edge of El Progreso, and passes between Lagunas Colorada and El Junco before eventually veering southward as it descends to its terminus at Puerto Chino on San Cristóbal's southern coast. Puerto Baquerizo Moreno is accessed from the ocean through entry into Wreck Bay,[3] which on any given day is populated by anchored fishing boats, pleasure yachts, water taxis, cargo and military vessels, and cruise ships that are boarded from separate passenger, commercial, and military docks. With a population of nearly 6,000, Puerto Baquerizo Moreno is the second-largest town in Galápagos and its center of government. A recently modernized airport accommodating daily flights with the mainland, Baltra, and Puerto Villamil, lies directly past the town's southern edge. A naval base dominates its western border along the southern shore of Wreck Bay. Town residents work in the numerous office buildings serving various levels of government and in the newly built hospital, the largest school on the island, and a university campus and science center on the town's northern edge at Playa Mann. Many residents are also directly or indirectly involved in the tourist industry, operate shops, hotels, and restaurants, or fish the local waters.

The small, sleepy town of El Progreso lies at 250 m asl, approximately 7 km along the main road into the highlands. Its focus is a large central park with a church at its east end, across the plaza from a health clinic,

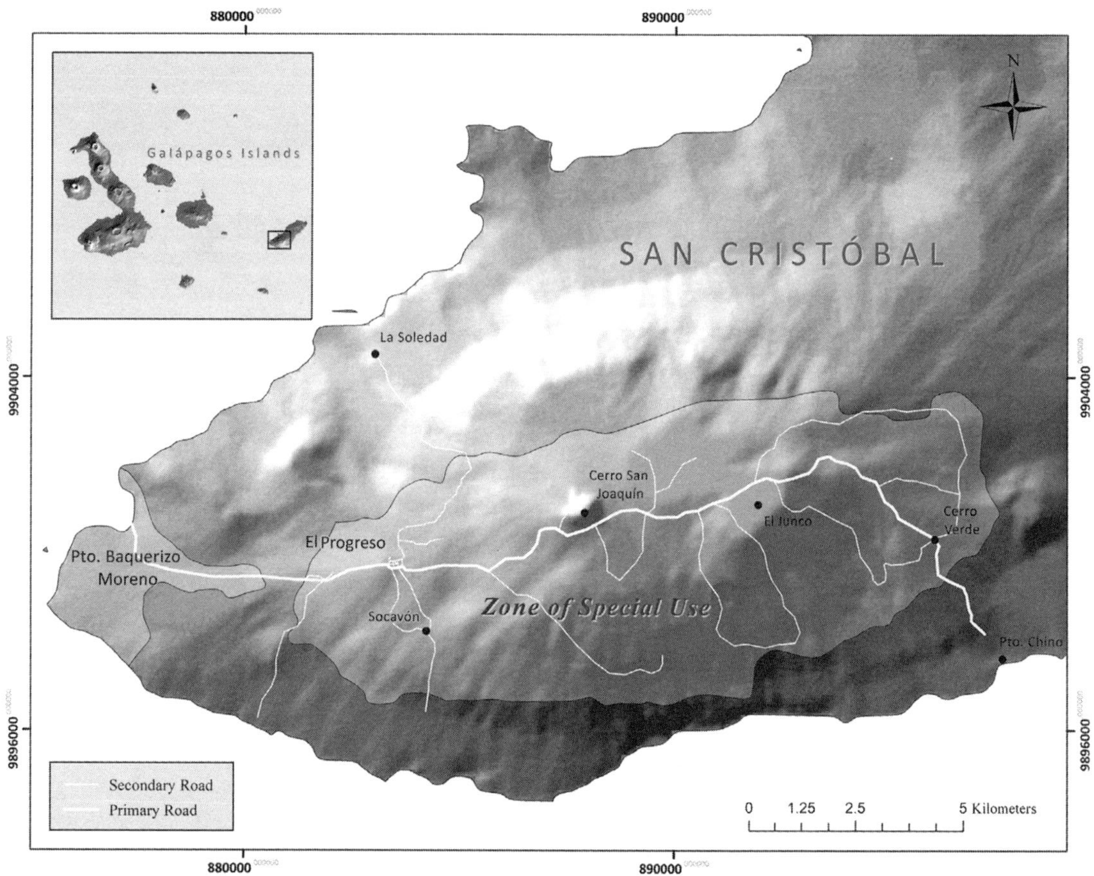

Figure 3. Southwestern end of San Cristóbal Island showing the populated areas and Zone of Special Use. Map Datum: UTM 15S WGS 1972, created by Rylee Harlos.

newly constructed police station, and a local government administration building on it western edge. Directly above these buildings, on the highest ground overlooking the ruins of the historic Cobos house site, is a water treatment plant where potable bottled water is pumped from a deep well drilled by the US military. Health offices and a roofed hippodrome are located on the southern edge of town where the main road continues inland. The Charles Darwin de El Progreso School[4] and enclosed playing fields lie directly to the south of the central plaza, where most of the 132 homes are located. Many of the town's almost 500 inhabitants also own property on the coast and agricultural lands in the interior, so many homes are often vacant. A few also serve as shops for basic supplies, or open as restaurants for noonday meals. Sundays are particularly lively, when families from the coast descend on El Progreso for football games at the school and lunch at the numerous restaurants that open solely for the weekend pastime.

The town services a large, 8,201 ha area divided into nine additional precincts, some with tiny hamlets, and all with variably sized farmsteads involved in gardening and ranching.[5] Principal crops include potatoes, maize, tomatoes, tree tomatoes, vegetables, yuca, plantains, bananas, watermelon, melons, pineapple, sugarcane, and citrus. An organic coffee plantation, El Cafetal, is located in Socavón, and recent initiatives involve greenhouse production of tomatoes, cucumbers, and peppers. Cattle, pigs, and horses are the island's principal livestock. Farm products are mainly consumed on a local level, and the lack of commercial production has resulted in significant abandonment of farmland for other pursuits. Problems include labor and material costs, low market prices, competition from imported products, poor organization and planning, and the presence of alien invasives.[6] Many of these, particularly guava, dominate the landscape above 150 m asl (Colinvaux and Schofield 1976: 992; Restrepo et al. 2012: 1855).

Manuel J. Cobos and Early Developments on San Cristóbal

Manuel Julián Cobos hailed from Cuenca, the largest city of Ecuador's southern sierra, where he was born in 1836. He subsequently emigrated to Chanduy, a southern coastal village in the country's western lowlands, and together with his brother Angel and fellow Cuencano José Monroy,[7] established Cobos y Hermanos, which by 1863 was one of the most prosperous businesses on the Santa Elena Peninsula. At this time both brothers faced accusations of capriciousness and violence, and despite assuming notable roles in local politics, both had run afoul of the law. Although Galápagos orchilla had likely been discovered earlier, serious enthusiasm for its economic potential began in 1860; from the outset, Cobos and Monroy, through their Empresa Industrial de Orchillana y de Pesca, installed provisioning operations on San Cristóbal for its collection.[8] By 1869, two sloops bound for Galápagos from Chanduy were exporting harvested orchilla, sea lion skins, and hides and salted meats from feral cattle herds on San Cristóbal and neighboring islands to Panamá, where higher prices were obtained. Upon their return, the sloops, appearing to have come directly from the islands, arrived in Chanduy laden with Central American contraband (Latorre 1991; 2002).

Thirty years into its history as an independent republic, Ecuador's public treasury was heavily reliant on export production generated in the western coastal lowlands, where its residents were reputed for their liberalism, independence, freedom from feudal domination, and interest in foreign trade. Unlike highland production, which was organized in extensive land holdings with a large indigenous workforce bound to estate and *patrón*, coastal plantations were more advanced and efficient, with better working conditions for laborers. Nineteenth-century expansion in world trade and demand for tropical crops prompted coastal production to focus on a restricted range of export commodities.[9] At this time, the bulk of Ecuador's national income was generated in the coastal lowlands, and since 1830, "customs, particularly import duties [were] the most important source of ordinary state revenue" (Rodríguez 1985: 53).

Cobos and his partner Monroy would again run afoul of the law. They faced charges in Guayaquil of smuggling contraband and of poaching cattle belonging to others. Also, the descendants of José Villamil, alarmed by the growth of Cobos's business and its invasion of their legal entitlement on Floreana, appealed to newly installed interim President Gabriel García Moreno, who in 1869 had also awarded the exclusive permit for collecting island orchilla to José Valdizán. In 1871, provincial and town authorities ordered the capture of Cobos and Monroy, and seizure of their assets. However, the partners, attracted by the 1870 discovery of orchilla in Baja California, managed to escape custody. Along with a labor force of 300 deserters and reprobates recruited by Monroy in Guayaquil, they descended on Magdalena Bay via Galápagos. Business operations on San

Cristóbal were left to a small, permanent labor force[10] charged with clearing land for pastures, domesticating cattle, and planting sugarcane.

Magdalena Bay, on the southwestern coast of Mexico's Baja peninsula, was the site of an earlier failed scheme by American speculators to colonize and develop agricultural and mineral lands during the 1860s.[11] Orchilla had been subsequently discovered on the property owned by the Lower California Company; however, in 1871 their conditional grant was declared null and void by Mexican President Juárez, in exchange for the exclusive 20-year privilege to collect orchilla under condition that subsequent land claims be abandoned and the government compensated 5 dollars per ton of exported product (Daily Evening Bulletin 1872a). Later court documents suggest that Cobos was involved in Baja orchilla collection as early as 1870, whereas earlier documents in a notorious court case claim that he had contracted with the Lower California Company, presumably after the September 1871 Mexican Supreme Court annulment, to gather orchilla and divide the profits equitably.[12]

Cobos was accused by company representative and US consul at Magdalena Drake De Kay, along with Mexican Collector of Customs Francisco Jiménez Mendizabal, of theft and running contraband. In the middle of the night on October 28, 1871, Cobos and Mexican officials "with a boat filled with armed Peons, from Equador" allegedly seized at gunpoint the American vessel *Cina Greenwood*. De Kay further alleged that Cobos had imported arms to equip some 200 men in an effort to capture and destroy the Customs House, which contained evidence of debt, kill Mendizabal, and seize control of the port to facilitate the unimpeded export of orchilla and import of contraband (De Kay 1871: 25). Documents countered that the *Cina Greenwood* was actually the property of Cobos and his partners, and a later visit by the US steamship *Saranac* corroborated that neither assault nor violence had been committed and that claims for revolution were unfounded. Cobos was acquitted on all charges.[13] In his own words, Cobos continued in Baja with his partners and 300 Ecuadorian laborers for nine years, until the Russo-Turkish War of 1878–79 paralyzed European manufacturing (Latorre 2002: 7).

Manuel J. Cobos and the Hacienda El Progreso, 1879–1904

Despite Cobos's claims about the Russo-Turkish War, the trajectory of his business pursuits was changing. The demand for plant-based orchilla was diminishing in the later nineteenth century, José Valdizán was assassinated in 1878, as was President García Moreno in 1875, and liberalism was emerging on the mainland.[14] With capital accrued from almost 10 years of business in Baja, Cobos and Monroy refocused their commercial ventures on San Cristóbal. In his own words, Cobos recounted that upon his return in May of 1879, the four-year-old sugarcane plantation was of medium size and producing successfully. Attempting to solicit funds from his old friends in Guayaquil, he promised an eventual daily production quota of 50,000 lbs. Despite his claims of return on their initial capital outlay within four years, potential investors were leery of investing their money in a setting of thieves and murderers (Latorre 1991: 9).

When Cobos arrived with a few of his Baja workers in 1879, San Cristóbal may have had as many as 150 human residents. Bognoly and Espinosa's (1905: 87) account of El Progreso after Cobos's death claims that with Valdizán's assassination, San Cristóbal's population had been tripled by the addition of 100 people who had fled Floreana. In January 1886 the resident Territorial Marshal reported a population of 148, which increased to 209 in August of that year (Guevara 2015: 56), and an 1887 visit to the island mentions 140 (Vidal Gormaz 1891: 160). Two decades later, Alexander Mann wrote that when Cobos returned, "at this period the island was teeming with cattle, horses, donkeys, and the gigantic tortoises" (Mann 1909: 28).[15] During an 1887 visit by the Chilean corvette *Chacabuco*, it was claimed that introduced animals, especially goats, were propagating admirably; however, the island was filled with many noxious animals like dogs, donkeys, and cattle. In 1889 the Territorial Marshal wrote how pigs and donkeys frustrated the establishment of gardens, and that enormous populations of dogs attacked young cattle and the few horses on the island (Vidal Gormaz 1891).

Cobos continued with his established business ventures during these early years in Galápagos. Although nearby Floreana had been totally abandoned by 1880, it was teeming with feral fruits and animals. El Progreso exported animal hides and oils, salted fish, cane alcohol, and molasses to the mainland, and exploited a wide variety of domesticated and wild products.[16] Although orchilla was still being harvested for export, it had become rare by 1887 as a result of incessant collection (Vidal Gormaz 1891).[17] Early on, Cobos focused his considerable energy on expanding cattle pasturage and increasing sugar production, both of which involved extensive land clearance and field preparation. An 1891 letter by the

Territorial Marshal mentions that for 12 years the sole occupation on the island was sugarcane cultivation and the operation of an animal-powered mill (Guevara 2015: 29).

Accommodations for sugar production were mainly carried out through human labor. The preparation of fields for cane planting was particularly laborious, as the removal of trees, shrubs, and volcanic rocks was undertaken largely by hand along with manual implements like hoes, forks, shovels, and machetes (Deerr 1911). After the new sections were cleared, heavier wood was harvested or pushed aside for use as fuel. Despite ample amounts of highland rainfall, sugarcane requires large quantities of water, so irrigation ditches were excavated. For ease of transport, fields were ideally situated around the central mill for cane processing, and roads wide enough for carts to pass between fields had to be constructed.

Colonel Pedro Jaramillo, the Territorial Marshal of Galápagos since 1885, reported that, not counting pasture, a total of 168 cuadras of land were under production in 1887, including 80 cuadras of rectangular fields[18] devoted to sugar cultivation, 30 to coffee and yuca each, 11 to vegetables, 6 to maize, 5 to potatoes, and 3 each to plantains and beans. In the following year, he estimated 370 cattle, 300 horses, 60 mules, 104 pigs, and 40 goats on Hacienda El Progreso (Latorre 2002: 15). By 1889, cane fields had increased to 200 ha (Latorre 1991) when Jaramillo estimated that San Cristóbal had a resident population of 287 inhabitants, comprising 213 men, 54 women, and 20 children. This number included government officials and soldiers, hacienda functionaries, 50 day laborers, and 55 indentured peons or *conciertos*.[19] Infrastructure included a road; five workshops; separate facilities (*ingenios*) for sugar and alcohol production; two stores, a warehouse, and an abattoir; two water reservoirs and field irrigation; 3 pastures and 17 garden sites (*chacras*); the hacienda main house and 60, mainly thatched, homes for government and hacienda workers; and 3 sloops, 2 smaller boats, a barge, and 4 scows in the bay at Puerto Chico (Vidal Gormaz 1891: 183).

Galápagos became part of coastal Guayas province in 1885, when the Ecuadorian government also decreed a series of inducements to stimulate island colonization. These included mortgage-free existence for 5 years, no military duty for island residents, and 20 ha of property; anyone wishing more was required to follow the regular procedures for acquiring vacant land. All duties were lifted, and any article could be imported to the islands free of duty and taxes as long as it remained on the islands.

Beginning in 1887, Cobos and Monroy began the process of importing state of the art machinery for sugar production from as far away as Scotland, and acquiring technicians for its installation on San Cristóbal (Bognoly and Espinosa 1905: 67; Latorre 1991; Vidal Gormaz 1891: 159).

Latorre (2002: 15) considers 1889 a decisive date in El Progreso's transformation from agricultural farm into industrial center and sugar plantation. During the final decades of the nineteenth century, Galápagos was attracting increasing international attention, particularly for its strategic position in the projected future of Pacific navigation and completion of a planned Panama Canal. The islands had also become a popular destination for various scientific expeditions that visited San Cristóbal during the hacienda's development into an island empire under Cobos's energetic direction.[20] However, our best descriptions of El Progreso at its peak under Manuel J. Cobos appear in accounts written during visits to the island after his death, in 1904 (Bognoly and Espinosa 1905; Webster 1904), 1905–1906 (Beck 1906; Slevin 1931), 1906 (Martínez 1915), and 1907 (Mann 1909).

The focal center of El Progreso was the hacienda house, and more precisely Cobos's panoptical second-floor bedroom from which he could inspect his empire in every direction. To the east lay the houses of his laborers, surrounded by extensive agricultural fields, and further inland the vast highland pastures dotted with cattle. Directly to the west were the sugar mill and processing buildings with views down the access road to the coast, Puerto Chico, and his boats anchored in Wreck Bay. The house, built on a promontory directly above the entrance to the modern town, was described by Martínez (1915: 45–46) as an unpainted and very ugly, yet ample and comfortable, fortress. The interior, which could be accessed on only one side, included a warehouse and stores on the first floor and, according to Latorre (2002: 36), narrow stairs leading to a second floor with a jail and bedrooms; Cobos's quarters were on the southeastern corner, from which he could monitor everything entering and exiting El Progeso. A 1904 plan shows an upper floor dominated by a central hall surrounded by at least a dozen rooms including a kitchen, dining room, hospital, galleries, offices, and Cobos's corner bedroom (Webster 1904: 137).

To the north of the house were a number of low buildings, including a warehouse, bakery, and store, which Martínez (1915: 39) tells us had all articles to be desired, including wines, liquors, and preserves. The *ingenio*, or processing plant and sugar mill, was the largest he had ever seen, including those on the coastal mainland. A large chimney dominated the zinc-roofed building, which housed imported machinery powered by three large boilers, each producing almost 112 kW of power, and fed by fuel from wood and bagasse, or the waste left over from pressed cane (Martínez 1915: 39).[21] Mann (1909: 29) describes the factory for elaborating sugar and spirits as "very well mounted and of a modern description," noting that the up-to-date machinery, principally from Glasgow, included "a large cane mill, triple effects, vacuum pan, centrifugal separators, and a number of other accessories."[22]

Gazing toward the island's interior highlands, one could distinguish the thatched roofs of the workers' quarters directly to the east of the house. Various photographs, a few with Cobos posing prominently, depict the simple huts against a backdrop of expansive fields in the distance. Martínez (1915: 46–48) called attention to the dilapidated government house[23] in the town's center, surrounded by productive cane fields and fruit orchards, which backed on to extensive pastures sloping up to the spreading base and rounded summit of a high mountain in the distance. On the northern edge of town lay the workers' gardens, and to the south was a coffee plantation, with a road wide enough to accommodate a rail line running through extensive cane fields lined by coffee bushes shaded by guava and plantain trees. Bognoly and Espinosa (1905: 166) estimate that more than 1,200 ha of the island was under production for the hacienda in 1904, including 1,000 cuadras (ca. 700 ha) of natural or unimproved pasture; 210 cuadras (147 ha) of foraging or improved pasture; and 622 cuadras (435 ha) of cultivated crops.[24]

Sugar was the focal enterprise at El Progreso; during his 1907 visit, Mann (1909: 29) claimed that 500 tons, in addition to alcohol, were produced annually.[25] Large-scale sugar production required heat and considerable inputs of water, fuel, and labor. Cane planting was a manual task, beginning with two or three 1-foot-long stem sections tamped by foot into shallow hoe-dug holes spaced 5 to 6 feet apart in long cane rows. During its 10- to 12-month maturation, the "plant cane" was weeded regularly by machete with dead leaves removed until stalks achieved sufficient height. At harvest, leaves on each of the 6 to 10 stems of each plant were stripped, and the cane was cut close to the ground, leaving enough stem and root for growth of the next "ratoon" crop (Deerr 1911: 123; Galloway 1989: 13–14; Mann 1909: 30–32).

The enormous volumes of water required to irrigate fields and supply the *ingenio* were transported from the interior highlands. Martínez (1915: 51) mentions that at one point Cobos had attempted to measure the depth of El Junco but stopped at 50 m. Both Martínez and Mann visited El Junco and mention its outlet as the waterfall at Freshwater Bay, and the springs and marsh on its western flank "from which the village is provided with drinking water by a canal cut in the ground where level, bridged by galvanised gutters where hollows occur, the crude aqueduct being over six thousand yards long, and carried down to the village by a gradual decline, on top of wooden posts, a miniature type of the ancient Roman aqueducts" (Mann 1909: 33). One source suggests that it was constructed at a cost of 35,000 sucres (Webster 1904: 10). Visitors to El Progreso after Cobos's death marveled at the perfection of the 6 to 7 km long system of zinc, iron, lime, and stone-lined conduits, canals, and pipes, installation of which began in 1898 under the skilled attention of Cobos's faithful Mexican assistant from his days in Baja, Felipe Lastre (Latorre 1991: 44–45; Webster 1904: 143).

A system of roads and trails were required to move products between field, *ingenio*, village, and coast. Martínez describes a good and wide road, proceeding some 5½ miles from Puerto Chico to El Progreso, and ascending a further 11 miles into the center of the island before branching into two 3-mile sections to the northeast and southeast (Latorre 1991: 43).[26] Harvested cane was transported from field to mill on a rail system using 50 cars pulled along 7 km of track by oxen, which Cobos had planned to replace with a steam locomotive (Bognoly and Espinosa 1905: 166; Martínez 1915: 47).[27] Transporting products up and down the steep climb between El Progreso and the coast was accomplished along the road by carts hitched to a team of two or three oxen (Martínez 1915: 47–48; Webster 1904: 10).

Visitors to El Progreso marveled at its bountiful groves of fruit trees, which Martínez (1915: 31) attributed to Cobos's introduction, yet they grew as if native to the island. They included orange, lemon, guavas that formed forests, plantain, ice cream bean, avocado, cherimoya, banana, hog plum, pomegranate, sapodilla, mango, and papaya, with apple, pear, peach, apricot, and plum grown in upper regions. Starting out with 30 plants in 1866, coffee was now propagated on a grand scale. Mann (1909: 29) provides an estimate of at least 100,000 "fine adult coffee trees" that bore two pounds per plant twice a year (Latorre 1991: 45).[28] Martínez (1915: 30–31) mentions that tobacco, introduced by Cobos with seeds from Havana, had since gone wild. Splendid crops of yuca were grown on a large scale, as were cocoyam, sweet potato, and beans. Experimentation with lentils, garbanzos, and peas had produced good results, and great quantities of exceptionally large ears of maize were harvested. Potatoes, barley, and wheat could be produced in colder areas, and pineapple, rose apple, watermelon, melon, cabbage, cauliflower, lettuce, turnips, artichokes, radish, squash, and gourds were cultivated. Martínez (1915: 29–30) attributed the introduction of *cabuya* (agave or century plant) to Cobos, and mentions an abandoned vineyard, destroyed by animals, which was once stocked with an Italian Moscatel grape varietal.

Riding inland from El Progreso in 1907, Mann (1909: 29) commented on the "wide extent of pasture land, almost devoid of trees and brush, which I estimate at over ten thousand acres" that extended to the mountain's summit at almost 2,700 feet. Noting that the soils were particularly suited for tree growth, the area was nevertheless devoid of trees and shrubs, yet nowhere did he find any evidence for their "artificial extinction," a phenomenon that he also noted in the highlands of Isabela. However, on San Cristóbal, wild cattle and land tortoises had been practically exterminated. A year earlier Martínez (1915: 33) advocated the introduction of fast-growing eucalyptus, pines, and cedar to replace the slow-growing native trees that would soon disappear through cultivation and wood use. In comparing the island pastures to those of his Andean homeland, he further suggested their improvement through introduced alfalfa, clover, cattle grass, and blue grass. Both visitors commented on the packs of completely wild dogs of various breeds, and the large numbers of colored and clear-skinned cattle, which were replaced at higher elevations by large herds of donkeys and mules (Mann 1909: 32; Martínez 1915: 49–50).[29]

Animal products, especially from cattle, factored heavily in El Progreso's internal and external economy since its earliest days. Latorre (1991: 46–49) reconstructs animal exploitation on San Cristóbal, as well as on Floreana, which was devoid of human population but teeming with feral animals.[30] Cattle and pigs, and to a lesser extent horses, were a constant source of wealth. Wild cattle on the island were re-tamed and kept in barb-enclosed pasture that excluded the large herds of donkeys and mules, which were domesticated for traction when needed.[31] Teams of hunters were regularly sent into the upper pastures to acquire cattle whose meat was consumed locally or dried as *charqui*. Along with meat from pigs and tortoise, it was exported duty-free together with hides obtained from cattle and sea lions. Sheep wool was also exploited, as mention is made of the death of El Progreso's shepherd, Liborio Escalante, from 200 lashes as punishment for losing some of his flock to wild dogs. Animal oil was also extracted from sea lions, marine turtles, land tortoises, shark, iguanas, sea bass, and even whales on San Cristóbal and other islands, especially Floreana and Isabela.

To the west lay El Progreso's terminus at Puerto Chico on the shore of Wreck Bay, backed by the Pacific Ocean and intermittent views of Santa Fé in the distant horizon. Martínez (1915: 43–45) and Mann (1909: 28–29) describe a 5–6-mile journey along a good road covered in the fine powder of ground red rock. Leading them to Puerto Chico through "extensive fields of sugar-cane" and upper forests with lianas, bromeliads, birds, butterflies, and the fresh humid scent of jungles, it descended through rock outcroppings, drier vegetation, and algarroba groves, with thousands of lizards crossing its path. In 1905, Puerto Chico consisted of a warehouse, plantation manager's house, a few simple homes, the Wreck Bay light (a lantern atop a lengthy bamboo pole) and the light keeper's house (Slevin 1931: 44; Webster 1904: 128). Martínez (1915: 43) mentions a small guard house, cabin for the ship's captain, and a vast warehouse for El Progreso's products.

Oil was extracted from marine mammals and reptiles on San Cristóbal's coast as well as from animals on other islands, especially Floreana and Isabela, where Cobos had long been sending his employees to procure feral products, especially cattle skins and meat. Sulfur was mined on Isabela, salt on Santiago, and lime only a short distance from Puerto Chico. Fresh fish was obtained for local consumption and dried for export to the mainland (Latorre 1991: 47–51).[32] Goods were loaded onto ships via a large 100 m long wooden pier equipped with a rail system (Martínez 1915: 23; Webster 1904: 128), and whose stone base can still be seen at the south end of the bay. Products were moved between the islands and mainland on two boats, the *Manuel J. Cobos* and *Josefina Cobos*, often the only available means for communication with the outside world.[33]

In 1905, Martínez (1915: 39) suggests that around 400 people[34] lived on San Cristóbal; however, we are certain that in the previous year, five days after Cobos was murdered, 90 men, women, and children fled the island on the *Josefina Cobos*, including German *ingenio* foreman Emilio Hansen, who was forced to pilot the boat (Webster 1904: 116). Various censuses undertaken by Territorial Marshals between 1886 and 1904 list a wide variety of occupations at El Progreso (Guevara 2015: table 8). Government employees had been resident on the island at least since 1885, when Pedro Jaramillo arrived in December of that year to assume the newly created position of Jefe Territorial. The National Congress also resolved to place a beacon in Wreck Bay and allocate 2,000 sucres to construct a building for the Territorial Marshal, his deputy, a secretary, and a schoolmaster (Bognoly and Espinosa 1905: 67–69). Other occupations included inspector, warden, guards, postmaster, and lighthouse keeper (Guevara 2015: table 8). Despite their presence, Martínez (1915: 37) observed that government authority was illusory; government representatives did nothing except to demonstrate the existence of governmental representation.

Effective power rested in the hands of Cobos, who was employer and patron for the island's inhabitants. The various occupations mentioned in the reports reflect the thriving industrial concern and self-contained community that El Progreso was in its heyday: administrator, bookkeeper, engineer, mechanics, sugar master, foremen, paymaster, office clerks, ship's captain, boatswains, pilot, sailors, carpenters, blacksmith, bakers, cobblers, cooks, belt-makers, tailors and seamstresses, sawyers, tin smiths, distillers, barbers, masons, launderers, and others. The bulk of El Progreso's workforce comprised day laborers, "colonists," "farmers," cattlemen, and fishermen. The August 1886 statistics list 67 colonists and 50 *conciertos*, together representing more than 65% of the population. Subsequent censuses variously refer to day laborers, *conciertos*, and farmers (Guevara 2015: table 8; Latorre 1991).

Much has been written about the composition of El Progreso's labor force. Many hacienda workers were criminals recruited from coastal jails, petty thieves and reprobates, debtors, military insubordinates, deserters, and unfortunate individuals secreted to Galápagos for a life of debt peonage. Some of the reported Peruvians and Colombians may have been undocumented and recent arrivals to Ecuador who were enticed to San Cristóbal with the prospect of work. Cobos absorbed his workers' debts and their incurred expenses with interest, keeping them in bondage by offering low wages in the form of private currency redeemable only on the island (Latorre 1991: 54–59). Bognoly and Espinosa (1905: 99) note that laborers were never paid in accepted currency or bank notes, but in rubber tokens, copper pieces, and paper bills.[35]

Life for island laborers was harsh. In their report, undertaken directly after Cobos's death, Bognoly and Espinosa (1905: 99–113) provide a detailed account of long hours, primitive conditions, little comfort, and severe punishment. The lengthy workday for field hands began at five in the morning and ended at six, or occasionally eight, in the evening. Work in the mill is said to have proceeded almost continuously from two in the morning until ten at night without shifts. Long days of up to 18 hours were reported for the office and store. Remuneration for day laborers was trivial, fluctuating between 8 and 14 reales per week, and paid in the hacienda currency.[36] For those who wished to partake, meals were provided by the hacienda at a daily price of 20 centavos.[37] Meals were prepared in different kettles; one, boiled unpeeled plantains with a repugnant color and disagreeable taste; the other was used for unacceptable fish that could not be discarded because the leftovers were fed to pigs. The village priest was a fellow peon, Pablo Quiñones, and religious affairs including marriages and baptisms were carried out with little ostentation. Long hours and poor pay kept the workforce in debt with little time for days off, which were provided only three times a year: Shrove Tuesday, Cobos's patron saint's day, and New Year. On these days, work stopped at two in the afternoon for a dance in the "repugnant" dance hall.[38]

Various conditions contributed to the dramatic events that surrounded Manuel J. Cobos's impending death at the hands of his workers. Martínez (1915: 40) and Mann (1909: 37) both emphasize chronic sexual imbalance as a primary cause for the brooding discontent at El Progreso. Estimates vary widely, some within the same year, but they all reflect a dearth of women among the resident population. Martínez claims that the proportion of women was not more than 15%, and for this reason polyandrous relations had become established.[39] Both visitors link sexual inquietude with alcohol and gambling, and an 1888 record of legal cases submitted by police inspector Vicente Jaramillo lists rape, sodomy, theft, and homicide in order of magnitude (Latorre 1991: 60). Foremost to blame for these conditions, according to Mann (1909: 37), was the unfortunate mixing of criminal elements within a population of otherwise respectable workers.

The authoritarian control exercised by Cobos and the harsh punishment he inflicted on his workers significantly exacerbated the stark conditions at El Progreso. Mann (1909: 34–35) spoke forcefully of the firm hand with which he held the community together, as he meted out severe corporal punishment for workers who did not fulfill their daily tasks or showed any signs of insubordination. Inhumane flogging, torture, and even murder "held his slaves in subjection," and resistance was hopeless on account of extensive espionage prompted by "the fear in which he was held." The few policemen on the island were entirely dependent upon Cobos and either condoned "his savage vagaries" or became "his unsalaried tools." Bognoly and Espinosa (1905: 112) related that Cobos had excellent secret police among certain female and peon confidants and the most rigorous sentence awaited any alleged conspirator. Punishment consisted of canings and lashings, with reports of 111 individuals having been whipped from 200 to 1,000 times, and 37 dying from their wounds. There were accounts of flogged and incarcerated workers who subsequently died of hunger or whose corpses were eaten by rats. Alleged conspirators were executed by firing squad, many simply disappeared, and others were exiled to uninhabited islands (Bognoly and Espinosa 1905: 105–110).[40]

Hacienda El Progreso after the Death of Manuel J. Cobos

On February 16, 1904, Colombian officials took custody of the schooner *Libertad*, which had appeared in Cabo Manglares under suspicious circumstances. Arriving from Chatham Island and bound for Mexico or Central America, it was seized with arms on board and no navigation papers. In the following days, telegrams were arriving in Ecuador that the schooner's passengers were being returned to Ecuador, eventually disembarking in Guayaquil on February 19. The detainees had confessed to the murder of Manuel J. Cobos and Leonardo Reina, after which they had absconded with the *Josefina Cobos* and 150 quintales of sugar (Webster 1904: 108–113).

The proximate cause of Cobos's death, from wounds inflicted by trusted employee Elias Puertas, a 30-year-old Colombian farmer, emanated from Cobos's refusal to mitigate a harsh sentence ordered by Reina. José Prieto was to receive 500 canings "and even shot" for attempting to set fire to a cane field, which in a later revelation was identified as a diversionary tactic in a plot to assassinate Cobos. Begging for clemency in the early morning hours of January 14, Puertas was harshly rebuked, after which he shot Cobos twice with the revolver that Cobos had lost days earlier. Escaping to his bedroom, Cobos armed himself with a rifle and from his balcony shot into the crowd below. As others broke through the door and penetrated his house, Cobos hurled himself off the balcony where he was further assaulted with shovels and cutlasses by the assembled crowd. Puertas had no idea who killed Reina, noting that many others were shooting guns at that moment (Webster 1904: 125–126, 141).[41]

The ultimate cause of Cobos's death is a matter of perspective. The immediate aftermath of the crime was met with great celebration amid shouts of "Viva la Libertad" and "Abajo la tiranía." The assembled crowd burned any books, papers, and documents from the hacienda offices and government house that were considered of importance. Both houses, the factory, and the store were ransacked. Furniture and personal items were removed from the government house, and 1,500 sucres were alleged to have been stolen from a suitcase in Cobos's possession. After the victims were buried, the conspirators had planned to stow sugar and sufficient water on the *Josefina Cobos* and leave for Central America with Hansen as captain. The remaining population would eventually depart with the arrival of the hacienda's other boat *Feliz Porvenir*. On January 20, 1904, 76 men, 8 women, and 4 boys left the island (Bognoly and Espinosa 1905: 124; Webster 1904: 116; 139–140).

For their part, the refugees told authorities that as the slaves of an inhumane master, they were tired of tolerating his many abuses. Working from year to year without pay, they only received punishment, torture, and beatings, and resolved to no longer live under these conditions. All Galapagueños knew of their 5 comrades who had died by firing squad, 6 who had succumbed to merciless whippings, one wretched woman who had been whipped harshly, and 15 who were exiled, some dying of hunger, on desert islands. The Territorial Marshal was complicit in their misery, as the authorities were under the domination of Cobos, the king of Galápagos and tyrant of Ecuador (Webster 1904: 113–114).

After a lengthy trial in Guayaquil the refugees were absolved of guilt due to conditions on the island. For their roles in the deaths of Cobos and Reina, respectively, Puertas and Panamanian thief Francisco Carranza received minimal prison sentences (Latorre 2002: 61). A recent assessment by Guevara (2015) considers the collective violence at El Progreso to have been a rebellion, conceived and organized in secrecy amid extreme control and repression. Both Mann (1909) and Martínez (1915) acknowledged that stark conditions on the hacienda were exacerbated by a chronic sexual imbalance and significant presence of criminal elements in the island population. Latorre (2002: 62) adds to this an absurd system of government and almost complete lack of justice and authority.[42] Alongside his apparent cruelty, Cobos is also remembered as an intelligent and visionary businessman of indomitable energy who created a small empire where previously there had been nothing (Latorre 2002: 62–63). Another assessment, written by his great grandson (Cobos 2000), treats most accounts as fables and lies meant to discredit a great man or to sell books biased by perspectives from the descendants of criminals. Instead, Manuel J. Cobos should be understood within the context of his time and the conditions of the islands in which he was one of the first permanent colonizers.[43]

Accompanying the first of three government commissions to investigate the El Progreso tragedy was the new Territorial Marshal Juan José Pino. Representing Cobos's business interests were José Monroy and civil engineer Arthur Reed, along with the refinery engineer, Daniel Campbell, who had buried the corpses three weeks earlier. In 1904, it was estimated that El Progreso was worth approximately half a million sucres,[44] and a census taken in April of that year tallied 180 men, 38 women, 11 boys, and 16 girls remaining on the island. At their first meeting in the hacienda house, Reed, who had been appointed hacienda administrator, only added to the anxiety of the island's workers by declaring that all incurred debts and obligations would not be pardoned, a task greatly hindered by the destruction of paperwork. Pino worked to improve conditions for workers by limiting working hours, instituting decent wages, facilitating access to subsistence gardens, and prohibiting the circulation of non-official currency (Bognoly and Espinosa 1905; Latorre 2002; Webster 1904). Conflicts between government policies and business interests culminated in a succession of officials and administrators, yet during both Martínez's and Mann's visits in 1906 and 1907, respectively, the hacienda was still producing at a high level, and population had expanded to 350, including independent farmers producing crops in as many as 34 *chacras* around the hacienda (Latorre 2002; Mann 1909; Martínez 1915).

The principal heir to the Cobos fortune was daughter Josefina, who at the time of her father's death was living in Guayaquil and married to Rogerio Alvarado. Assuming control of El Progreso in 1909, Alvarado immediately announced ambitious plans to colonize other islands, introduced his own currency, and laid claim to the entire island, as the hacienda was being encroached upon by ex-peons and colonists. Alvarado's business dealings quickly placed El Progreso into debt with mainland banks and Guayaquil businessman Lorenzo Tous. It continued to function until 1917, when usable equipment was sold to the Hacienda Los Alamos in Milagro, with the remainder falling into decay (Latorre 2002: 68–69; 2011: 222). In 1918, an illegitimate son, Manuel Augusto Cobos, arrived from France and was given a small section of the hacienda. A visitor to El Progreso in 1919 described 3,500 head of cattle, horses, patches of sugarcane, coffee, limes, and "a few hundred acres under cultivation where there ought to be thousands, and two hundred bone-lazy peons do the work of fifty ordinary farm hands" (Stock 1921: 120–121). The following year, another visitor observed miserable huts amid squalor, filth, heaps of rubbish, and swarms of fowl and dogs (Muhlhauser 1924).

Over the objections of Alvarado, the government conceded thousands of previously uncultivated hectares of land around Stephen's Bay and Puerto Grande on the island's western side to a failed colonization scheme in 1922 (Latorre 2002: 69–70).[45] During the decade, Galápagos experienced various Norwegian settlements on Floreana, Santa Cruz, and San Cristóbal, for which El Progreso served as a crucial provisioning base. Early Norwegian adventurers to the islands were guests of Cobos in January 1922, when he claimed that his fully functioning refinery was still producing 15,000 kg of sugar monthly. It was estimated that approximately 200 inhabitants were on the island at the time, in addition to 1,000 acres of cultivated land and 10,000 freely roaming cattle (Hoff 1985).

In August 1925, Norwegians aboard the *Floreana*, loaded with supplies for the colony on Floreana, were greeted by Cobos and Alvarado on the beach at Wreck Bay in a Ford pickup truck. Invited to El Progreso, they encountered a nature "much more like the Norwegians expected" with orchard groves, vegetable gardens, grazing cattle and horses, a "sea of sugarcane," dogs and chickens everywhere, and electrical wires strung between the houses and factory. Offered a tour of the *ingenio* by Cobos, they described it as "dilapidated and badly in need of repair" (Hoff 1985).[46] El Progreso was also visited in 1925 by French tennis star Alain Gerbault, who dined with 25-year-old Manuel A. Cobos. Skirting "various ramshackle buildings, a sugar refinery, and a saw mill" he reached the hacienda, in front of which was the semicircle of thatched huts housing the peons. Struck by the amazing quantity and astonishing variety of firearms and armed servants in the house, Gerbault was regaled during dinner about the dangers of life in El Progreso.[47] Pointing out that only one-tenth of the fertile island was under cultivation, Cobos also lamented that the inability to procure labor and a lack of communication were retarding the colony's development (Gerbault 1929: 62–66).

In September 1926, Alvarado and Cobos negotiated the settlement of Campo Noruego, only to see it largely disintegrate within less than four years. Only 14 remained when San Cristóbal was visited in 1928 by British adventurer William Robinson aboard his ketch *Svaap*. After finding a few "bleak, deserted" shacks on the beach, he made his way to El Progreso on a powdery red earth bullock road. Robinson described a "small, decrepit sugar mill with its tottering black chimney," frame shack government house, and a village with 200 peons whose squalid huts were laid out in three rows with "only mud and a little grass," towered over by Cobos's stone and wood house. After spending the night amid swarms of mosquitoes and "legions of rats," he recounted riding on horseback in a world of *vaqueros*, peons, herds of feral animals, coffee, bananas, and burning fields of sugarcane (Robinson 1932: 52–59).[48] In the same year, El Progreso was visited by American politician Gifford Pinchot, who reported that the sugar mill and truck were "temporarily out of commission." Pinchot had lunch in an old stone house, whose appearance harkened back to better days and suggested that it served as both dwelling and fort. Afterward, he rode with Cobos through fruit groves, coffee plantations, rich grazing uplands, and "under forests of guava trees whose delicious golden fruits literally littered the trail" (Pinchot 1930: 84).

By the time Robinson revisited El Progreso aboard the *Svaap* in 1934, Cobos explained that the hacienda had been lost to a company which now owned it as well as the wild cattle, horses, coffee plantation, and schooner, he having been able to salvage "only a small plantation for himself" (Robinson 1936: 188). The hacienda and its infrastructure were in a continuing state of decay, when by the 1930s its control had passed into the hands of Tous, owing to Alvarado's mounting arrears, reported to be as much as 400,000 sucres. In order to absolve himself of debt, Alvarado accepted a proposition by Tous to divide the estate into shares as part of the "Sociedad Nacional Galápagos," retaining 25% for himself with controlling interest assumed by Tous. Alvarado subsequently departed for Guayaquil, and after a brief absence Cobos returned to the island, retaining the position of manager for the 300 inhabitants who depended upon it for their livelihood (Latorre 2002: 71; Otterman 1985: 152–153). Luís Guevara Traveso, who left his position as hacienda administrator in 1931, provided the following description:

> the state of the sugar plantation is disastrous; it has not worked for years; it needs a complete renovation that would require an expenditure of 30 or 40,000 sucres. Most of the fields are lost, filled with guava or bad weeds, which in order to recuperate the plantation would necessitate renovation of the sugar cane fields. The sugar that is consumed in the island is taken to Guayaquil. There are cattle in the hacienda and the island, but they almost do not produce sufficient milk for the population. The coffee plantations are invaded by weeds and have to be pruned and renovated. (Latorre 2002: 71)

Under increasing international pressure from naturalists, the short-lived presidency of Federico Páez Chiriboga created The Wildlife Sanctuary of the Galápagos Islands, turning Galápagos into exclusive property of the state in 1936.[49] Those with prior land claims were obligated to demonstrate proof of legal ownership; after a successful petition, Alvarado and his three sons were awarded 2,770 ha of hacienda property. This was to be used as debt repayment to the Sociedad Nacional Galápagos; however, subsequent litigation prolonged the transaction. Meanwhile, Tous planned improvements, bringing potable water to the port and fixing its decaying pier (Latorre 2002: 72).

Ownership of the hacienda eventually passed into the possession of Tous in 1938 when an official decree clarified that El Progreso village, the lands cultivated by settlers, and all feral animals were excluded from the sale. Abandoning sugar production, the society subsequently transported usable machinery to the continent, leaving large installations in place to decay. The neglected fields were quickly overrun with cattle, pigs, horses, donkeys, and guava as business interests shifted to fishing and coffee production (Egnal 2015). A 1938 visit to El Progreso briefly mentions the very primitive, garbage-strewn village, whose street was a sea of mud during the rainy season. It included a church with a skeleton of a tower and one large store, the town meeting place, which was overrun by animals and where dogs and cats also "live, love, and fulfill every function of their existence." Nearby was the wreckage of the factory with rusted and desolate boilers sticking out of the tangled vegetation, the Norwegian houses in poor repair, and the home of Cobos, his wife Karin, and their four children (Seligman 1947: 285–290).[50]

During the war years, the focus of island activity had shifted to Base Beta, the American military installation on Baltra. Cobos, like many other residents, was employed there in the war effort. Alvarado died in poverty on the mainland, having separated from his wife and squandered the family fortune. San Cristóbal was important for its potable water, which supplied the base; otherwise, the island was isolated, having been basically forgotten by the government.[51] The society's remote administrator was unable to communicate with the mainland for assistance, and the hacienda was practically abandoned. After the war, cattle ranching and coffee cultivation continued under Tous's management as "La Predial," joining with American Mike Mann to purchase a fishing fleet and build a freezing plant in Puerto Baquerizo Moreno through an overall capital outlay of 20 million sucres. The freezer was completed in 1950, but the project failed when conflicts between the principals caused Mann to pull out of the fishing venture (Egnal 2015; Latorre 2002: 73–74).[52]

By 1960, migration from the mainland had augmented the population of Galápagos to 1,900, with some 1,200 residents on San Cristóbal. As early as 1958, small advertisements soliciting 50 adventurous families "to establish a model community on a beautiful Pacific Island" began to circulate in the Pacific Northwest. Eventually 106 members, including 22 families and 16 single men, committed shares of $2,500 to the Filiate Society Antrorse,[53] a colonization scheme conceived by Don Harrsch, a logger and tugboat operator from Washington State. Although Harrsch had never visited Galápagos, his research had targeted the islands as a profitable venue in which to locate a communal cooperative built on coffee, cattle, lobster, seaweed, tourism, and biological research (Faris et al. 1964).

On March 16, 1960, 30 colonists arrived in Wreck Bay aboard an old tuna clipper, the *Alert*, which had been purchased by the society during the previous year. They proposed to purchase Tous's abandoned freezing plant on the outskirts of Puerto Baquerizo Moreno as well as the 64,000 acre El Progreso hacienda. Their original plans failed almost immediately. Expected cash income from lobster and tuna fishing was not realized, and possession of the coffee plantation was obstructed by the lack of a clear title. Attempts to raise crops failed miserably.[54] Anti-imperialistic hostilities on the mainland delayed the arrival of a second group of colonists aboard the converted refrigerator ship *Western Trader*, which also had to leave behind vital supplies because of a 100% duty demanded by the Ecuadorian government. By January 1961, all but one of the colonists had departed from San Cristóbal (Faris et al. 1964).

Today, the footprint of Hacienda El Progreso defines the large Zone of Special Use located in the highlands and southern tip of San Cristóbal Island. What was once the historic property of an industrial-scale plantation and cattle ranch is now home to a patchwork of small land parcels, mostly privately owned by contemporary families, many descended from the island's earliest inhabitants. They continue to live and farm in an island interior whose landscape has been thoroughly transformed over the centuries through human activity. Immovable infrastructure from the hacienda's past has been left in place to decay, while many of the more conspicuous and portable objects have been relocated to various inhabited parts of the island, primarily for decoration. The remainder of El Progreso's preserved material culture lies scattered about the landscape and buried underneath homes and gardens.

4

Island Production for Global and Local Consumption

With capital accrued from a decadal sojourn in California, and accompanied by some of his loyal Baja workers, Manuel J. Cobos returned to San Cristóbal in 1879 and immediately focused his considerable energies on sugar production. Although the medium-sized plantation was already producing successfully, sugarcane cultivation and the operation of a mill became the island's sole occupation for the next 12 years (Guevara Ruiz 2015: 29). Cobos's return coincided with the global demise of the orchilla market and the 1875 assassination of President García Moreno, a conservative highland autocrat despised by the liberal coastal elites. His death ushered in a series of interim governments, military rule, restoration, and eventually the Liberal Revolution of 1895 under Eloy Alfaro which was to govern the republic for 30 years.

For most of the nineteenth century, Ecuador was the world's largest producer of cocoa. The "*pepa de oro*," grown in coastal haciendas since well before independence, fueled the earliest of a series of recurring export booms that have characterized the country's economic fortunes for centuries (Bromley 1981: 20; Redclift 1978: 40–41; Rodríguez 1985: 22, 99; Schodt 1987: 26, 36). In addition to cocoa, which constituted more than half of Ecuador's exports, other coastal lowland export products included tagua (ivory nuts), rubber, Panama hats, tobacco, cattle, and hides (Church 1983: 13). Although sugarcane was produced in various parts of the country, its consumption was primarily local (Geerlings 1912). Interestingly, sugar first appears as an export commodity of any significance between the years 1879 and 1904, when it rarely exceeded 3% of total exports by value (Rodríguez 1985: appendix A, 178). Unlike its highland counterpart, coastal agrobusiness was logistically conducive to foreign trade and export production. Coastal producers were well positioned to supply an increasing world demand for tropical crops and expanding global markets (Redclift 1978: 39; Rodríguez 1985: 14; Schodt 1987: 37) amid heightened awareness of widening possibilities prompted by the anticipated completion of a Panama Canal (Church 1983: 34).

Sugar production had an early start in New Spain with the establishment of Hernán Cortés's hacienda at Tuxtla in the early decades of the sixteenth century (Barrett 1970). By the time Cobos initiated his Galápagos *ingenio*,[1] Cuba had become the world's leading sugar producer, responsible for one-third of global supply (Tezanos Toral 2015).[2] Until the application of steam power, sugar production had improved very little from seventeenth-century practice based on fourteenth-century techniques developed in the Mediterranean sugar industry (Barrett 1965: 148–149). Early production relied on the motive power of oxen, which in turn required pasturage and the importation of guinea grass (*Panicum maximum*) (Moreno Fraginals 1976: 99–100); indeed, since its earliest beginnings in Mexico, the *ingenio* was intimately associated with raising livestock, especially cattle and mules (Barrett 1970). However, as Tezanos Toral (2015: 78) emphasizes, "one of the features of the Cuban sugar planter was his constant disposition to invest in new machinery and new methods of making sugar, no matter the price," a pattern apparently followed by Cobos in his remote Galápagos operations.

From its start in the 1860s, through expansion beginning in 1879, to its apogee in 1904, and continuing through its various later manifestations, the legacy of Hacienda El Progreso still survives in the highlands of San Cristóbal and the small town that bears its name. The ruins of its buildings and machine parts are scattered about the southern end of the island, the accumulated refuse of its productive activities is preserved in village soil, the imprint of its landscape transformations defines a Zone of Special Use that continues to be farmed, and its history lingers today in scattered written accounts, and in the memories of its descendants.

Archaeological Investigations of Hacienda El Progreso

Traveling eastward from the coast, the road ascends into the highlands and passes through the entrance to El Progreso, marked by El Rondel, a traffic circle that accesses three entrances into town. The small island in the middle of the circle appropriately displays large gears and rollers from an old cane press, set in concrete. Bearing left, the upper road continues to climb a promontory as it passes below the historic hacienda's commanding house site before entering the town center. Lined on either side by the potable water plant, junta parroquial, police station, and a few homes, dominated by the town park with the church at its far end, the upper road continues its journey out of town toward La Soledad. The middle road passes by the town carpentry directly below the hacienda house, and El Ceibo (claimed to be the world's tallest ceiba tree) across the street, before it accesses private residences and dead ends at the town soccer field across from the church and adjacent to the school. The lower southern road passes by Cobos's tomb, various residences, and a few small shops before it resumes its upward climb toward El Junco, after which it eventually descends to its terminus at Puerto Chino on the island's southern coast.

After a preliminary visit to San Cristóbal Island in 2012, we began our interdisciplinary archaeological investigation of the historic Hacienda El Progreso in 2014, which we continued each summer through 2018. The Historical Ecology of the Galápagos Islands project assembled Ecuadorian and Canadian scholars and students in a program of ongoing excavation, analysis, and community engagement centered in the town of El Progreso. The project's goal was to explore the spatial and temporal depth of human ecological transformation on San Cristóbal Island, and to elucidate the complex and historically contingent development of its novel anthropogenic ecosystem through the lens of historical and archaeological data.

Activities in 2014 included town meetings, preliminary reconnaissance, field sampling of soil columns for phytolith extraction, mapping of the town site, and brief inspections of the properties on which the old sugar mill was located (Plate 1). Excavations were undertaken adjacent to an abandoned historic structure locally referred to as La Cárcel, and in the principal historic midden located beneath the town carpentry. During 2015, excavations continued in the main midden, road cut profiles were cleaned to gauge the midden's extent, test pits were placed within the town site, and 20 units were opened throughout the central area of the town center, in the abattoir, and on the historic hacienda house site. In addition to continued mapping, repeat photography was undertaken from precise locations matched to perspectives of historic photographs, and analysis of recovered artifacts and organic materials began in Puerto Baquerizo Moreno. At the end of the season, shallow excavations below the historic house site next to the sugar mill area revealed the beginnings of a cobble pavement. The latter was extended in excavations during 2016, along with continued analyses in Puerto Baquerizo Moreno. Analysis and report writing continued in 2017, with further inspection of the sugar processing area after permission was obtained at the end of the field season to access two specific properties. This area was mapped with aerial and terrestrial LiDAR in 2018 (Plates 2 and 3).

The Hacienda House

The historic hacienda house site is today a domineering focal point of El Progreso. Situated atop a promontory, it looms over the western edge of town where it was strategically placed to overlook the sugar mill directly below and to observe approaching traffic with its panoramic view to the coast.[3] To the east, this location afforded a commanding view of the village against a backdrop of distant fields and pastures rising into the highlands (see Repeat Photos in Plates 4–11). A lone towering coconut palm believed planted by Cobos himself readily identifies the house site from a distance. Diagonally across town to the southeast stands another towering coconut palm, which marks the current residence of one of his descendants.

Some historic photos of the original Cobos house exist; two from 1888 (Figure 4) depict a massive and sprawling building, supporting Martínez's (1915: 45–46) description of an unpainted and very ugly, yet ample and comfortable fortress. The house site today (Figure 5) consists of a large stone base foundation supporting ornamented plastered concrete walls on its southern and eastern exposures, the latter having suffered partial collapse early in 2015. The surviving superstructure is clearly from a later construction. Exposed inner walls on its southern flank reveal concrete reinforced with what appear to be the rails from the small gauge Decauville system used to haul cane from the fields to the mill. Visitors in 1938

Figure 4. El Progreso hacienda house in 1888. *Left*: "Senor Cobos' House, Chatham Island, Galapagos" 1888. (View NE-SE). National Archives photo no. 22-FA-90 (Albatross Expedition). *Right*: "House of [?] Cobos, Chatham Island, Galapagos Archipelago" 1888. (View E-W). National Archives and Records Administration photo no. NARA 22-FA-98 (Albatross Expedition).

described a new house owned by Lorenzo Tous that was constructed on the original site. A 1947 photograph clearly shows a frame house incorporating the plastered walls under construction directly on top of the larger stone foundation that survives today.

The massive, up to 17 m long, rough stone foundation leveled the eastern slope of the knoll and may have supported the original house.

Figure 5. El Progreso hacienda house in 2014. (View E-W).

Constructed of local volcanic rocks cemented with earth and lime, it is almost 2 m high in places and surmounted by a larger built-in staircase on the eastern side of the structure, with two smaller staircases to the north and south. The lack of an upper façade entrance, to match the main staircase, further suggests that the stone foundations were built for an earlier structure. A surviving interior staircase comprising seven cement steps may have been associated with a later structure (Bolaños et al. 2002).

Touristic improvements were undertaken by a Spanish mission to San Cristóbal in 1999. The house site was surveyed in anticipation of a future interpretive center that never materialized; however, some restorative efforts, stone-lined paths, and signage, were completed. Smaller structures, not necessarily associated with the historic hacienda house are found in the immediate area. A shallow 2.5 m^2 water retention pool, consisting of a 25 cm thick rock wall bound in mortar and faced in cement, is located to the northwest of the house. The approximately half meter deep pool is accessed on one side by three interior steps. Two outlooks, based on large volcanic rock outcrops that were modified at some time by the addition of smaller rock stairs, and reinforced with cement and marine sand and shells, supply eastern and western views immediately to the north of the house structure. An approximately 4 m^2 rock feature, with a possible brick chimney base and 3 m diameter semicircular footing to support a dome on the north side of the outlooks, has been referred to as an oven.[4]

Earlier surface collecting in the area of the hacienda house recovered fragments of porcelain and glass, metal bolts, nails, domestic mills, couplings, cleats, axes, and parts of bridles and ornamental grates, among other items (Bolaños et al. 2002). In 2015, we mapped portions of the historic hacienda house site, and we excavated eight 1 m^2 units in the area. Although the historic house site is currently private property, the surviving features are subject to federal cultural heritage laws and in 2017 the process was begun to officially recognize this and another nearby area as a national historic site.[5]

The Sugar Mill

The site of the historic hacienda's sugar operations lie directly to the west of the house site below a steep incline to the north side of the road, immediately before it enters town. Most of the surviving sugar mill equipment and preserved infrastructure are found today on two linear tracts of private property, separated from the house site by a strip of public land which preserves the old public right-of-way at the base of the hill behind the town bus stop. In 2014 we excavated a test pit next to one of the standing buildings for phytolith recovery and took photographs of the sugar mill site. After a protracted setback, we were eventually granted permission to enter the properties again at the end of the 2017 field season.[6]

Carpintero Midden

The residence and workshop of the town carpenter are located at the entrance to town on a point between the upper and middle roads, directly below and across the street from the hacienda house site. Access to the workshop is from the upper road, where it is perched on a thin strip of land that plunges precipitously to the family residence and small courtyard below. The sudden drop between the workshop and the floor of the courtyard reveals a continuous earthen profile along the property and exposes the historic hacienda's midden. Four to five horizons are variably visible throughout its length (Figure 6), including: an active soil layer; a thick clay-rich, spheroidal- to euhedral-clod A horizon with various dead roots and some charcoal and cultural material; a cultural midden appearing as a buried darker mineral horizon composed of a solid wet clay matrix, densely packed with bone, charcoal, metal, glass, and pottery; a wet clay stratum with carbon flecks; and, a sterile base of clay saprolite. The soil averages 7.7 pH (Madelyn Percy 2015, pers. comm.).

Although the midden can be traced for some 20 m along the exposed profile, only a portion of the western half of the profile behind the house and courtyard and adjacent to the pilings that support the workshop directly above could be excavated (Figure 7).[7] Using the exposed profile as a guide, we excavated each depositional stratum in arbitrary 10 cm levels whenever possible. Excavations variably cut as much as 1 m or more of soil from the exposed face directly below the narrow walkway to the workshop above. In the course of exposing and removing the midden, we excavated roughly 10 m^3 of soil. In 2015, at the request of the carpenter, we similarly excavated directly behind the house adjacent to the 2014 excavations, removing less than 2 m^3 of soil. We screened the excavated deposits through 6.35 mm screen; however, this was often difficult due to

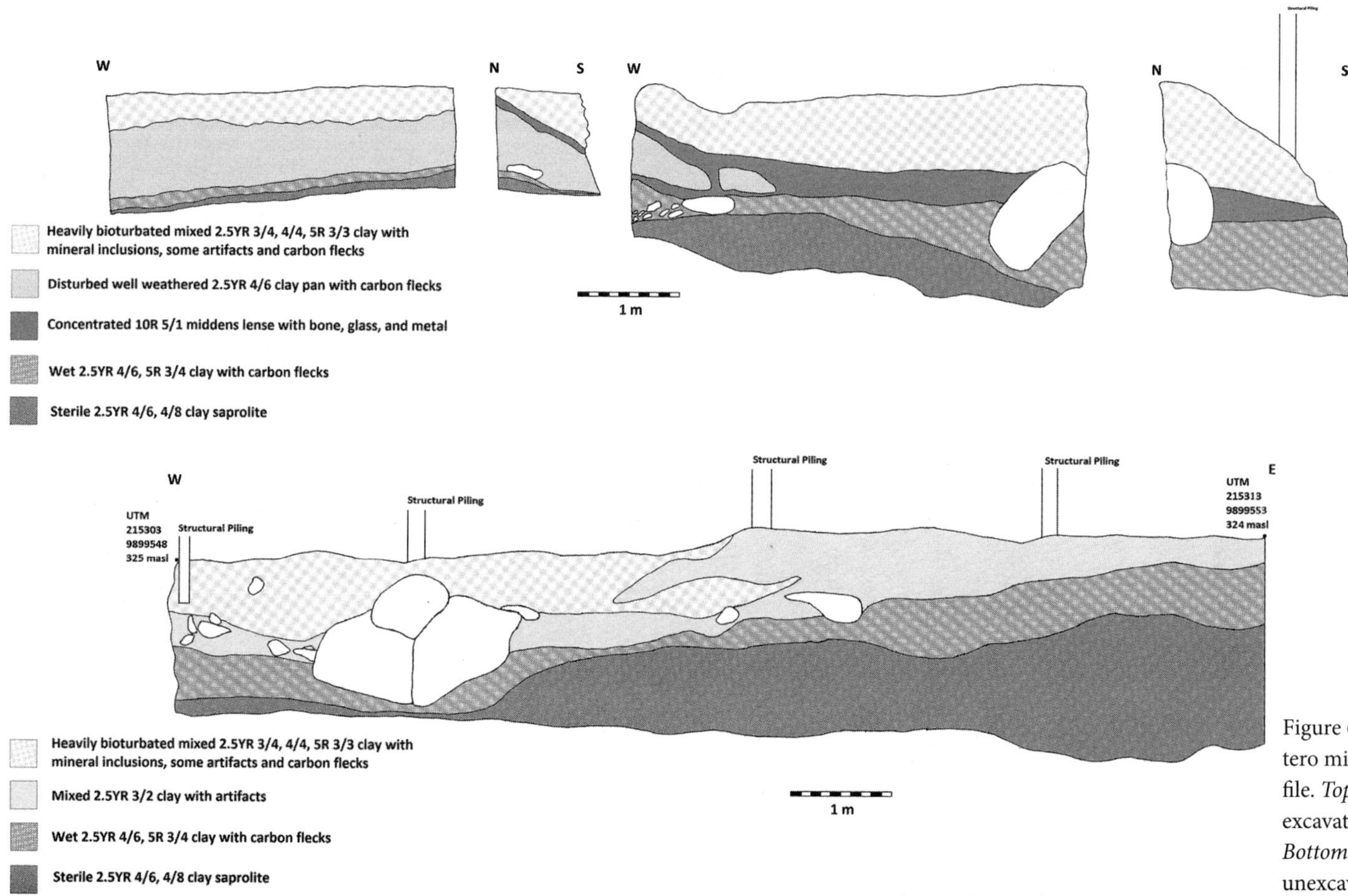

Figure 6. Carpintero midden profile. *Top*: western excavated half. *Bottom*: eastern unexcavated half.

Figure 7. Carpintero midden. Central portion of western half during excavation.

persistent showers and perpetually wet conditions. Because of the size and density of artifacts in the midden lens, much of the sample was recovered by hand.

We cleaned sections of the exposed upper road cut profile directly across from the workshop below the hacienda house site, and inspected the property down the slope from the carpenter's residence on the other side of the middle road, yet found no trace of the buried midden. We suspect that this was the historic garbage dump for both the hacienda house and the communal kitchen, which historic photographs show below the house on the western edge of the village. Refuse disposal appears to have formed a sheet midden down the slope from the hacienda house and communal kitchen, much of which was subsequently obliterated during later building construction and roadwork along the southward sloping entrance to town.

La Cárcel

Directly across from the historic house site, on the south side of the main street, are the ruins of a two-story structure, regularly choked with vines and vegetation, and locally referred to as the Hacienda El Progreso's jail

Figure 8. La Cárcel in 2014. Excavation units were placed in the courtyard on the other side of the peaked wall.

(Figure 8).[8] In the period between our 2012 and 2014 visits, the upper façade collapsed entirely; an earlier photograph of the cleared edifice taken only a decade earlier (Bolaños et al. 2002) shows an even more complete structure. The main entrance faces the street, passing through a 6 cm thick concrete and sand wall buttressed by regularly spaced upright wooden (matazarnos, *Piscidia carthagenensis*) beams supporting the upper floor joists. The interior wall surfaces show that the exterior concrete face was backed by horizontal laths of bamboo and reinforced with barbed wire. The upper story was accessed by a staircase resting atop a squared base across from a secondary door opening on the structure's east side. A number of vertical openings served as windows for the second story, whose floor, having been described as constructed of cement and lime, may have survived until recently (Bolaños et al. 2002).

Whether or not the structure served as a jail, or perhaps more likely as the store/warehouse of Lorenzo Tous's later ventures, the property is littered with old refuse. The area may have been the location of a communal kitchen, which historic photos place directly below the hacienda house. While clearing the area, Bolaños et al. (2002) earlier recovered fragments of a domestic mill, bolts, nails, various iron objects, and a large quantity of empty glass liquor bottles. Over the years, we had the site maintained, and in 2014 after clearing it of vegetation and summoning the town garbage truck to haul away years of accumulated garbage, we excavated two 2 m^2 units (Units 1 and 2) with one extension (Unit 3) on the east side of the property which was covered by ornamental plants, several types of grasses, herbs, bananas, and palms. Excavation revealed an active 8 cm thick soil layer, a 12 cm modern cultural layer, a clay A horizon with dead roots, charcoal, and cultural material possibly associated with the hacienda, and a mineral horizon composed of solid red clay.

Town Site Excavations

A series of excavation units and test pits were excavated during the 2015 field season in various areas of town. A number of units were spaced throughout the central park (Park Units 1–5), which was scheduled for resurfacing with brick walks and construction of new public facilities completed the following year. Test pits were excavated in a vacant lot on the south side of the street across from the town park. Additional excavation units were placed in the area of the town abattoir and in the courtyard of a residence below (Camal Units 10–12). Historic photographs depict this area of the modern town as the site of the thatched huts and a few wooden buildings that housed the workers throughout the hacienda's productive years.

Archaeobotanical Sampling

Various contexts were sampled for the recovery of archaeobotanical materials, including charred wood, macrobotanical specimens, and phytoliths. Standard 10 L bulk soil samples were collected from midden and test pit contexts. Wood charcoal was hand collected from exposed profiles during excavation and from light fractions of flotation samples.[9] Soil samples for phytolith extraction were also collected in columns from exposed excavation profiles, and from the profile exposed in the Sugar Mill area. The latter, excavated next to one of the standing buildings in an area covered by grasses and banana trees, revealed several layers of charcoal and ash.

Four 60 × 60 cm test pits were excavated to a maximum depth of 60 cm for placement of 15 × 60 cm columns in order to sample soils from the last 200 years in village, abandoned field, forest, and cultivated field contexts of the agricultural area in the Scalesia Zone. One column (CLM 1_VILLAGE) was extracted from the south profile of an archaeological test pit excavated in the rear yard of La Cárcel (0°54'27"S 89°33'27"W). The test pit was excavated in an area that may have been the location of the hacienda's communal kitchen. Four horizons were observed, including a buried cultural horizon possibly from the hacienda period. A second column (CLM_2 ABANDONED FIELD), was placed in the north profile of a test pit excavated to the southeast of the urban center (0°54'57"S 89°33'9"W) in a location covered by secondary forest and introduced plum rose (*Syzygium jambos*) resulting from the abandonment of a former agricultural field. A third column (CLM_3 FOREST) was extracted from the north profile of a test pit excavated near the border between the agricultural zone and the National Park boundary (0°53'20"S 9°32'48"W). The area was dominated by several native and exotic shrubs and small trees (*Bursera graveolens, Zanthoxylum fagara, Cordia lutea*; the native herb *Rhynchospora nervosa* was also present; and introduced *Psidium guajaba* and *Citrus sinensis*). A fourth column (CLM_4 CULTIVATED FIELD) was placed in the north profile of a test pit excavated in a modern cultivated field (0°54'6"S 89°33'55"W). This location is currently a fallow field in which corn (*Zea mays*) and fruit were grown, and is now covered in panicoid and chloridoid grasses. It was probably a sugarcane field in the late nineteenth century.[10]

The Archaeological Legacy of Sugar Production at Hacienda El Progreso

Visitors to El Progreso in the years directly after Cobos's death described the large mill and its modern machinery (Figure 9),[11] which eventually suffered rapid and chronic decay. As early as 1917 usable equipment was sold to mainland interests, residual on-site installations fell into disuse, materials were locally repurposed, and bits and pieces were regularly carted away as curios.[12] The mill area directly below the hacienda house is today littered with the surviving vestiges of earlier sugar production (Plates 1–3). These include water control features, stone and mortar structures enclosing furnaces, the possible base for the large chimney that dominated the *ingenio*, foundations and hardware from smaller buildings, wall stubs, and two large metal boilers.

Plate 1. Aerial overview of El Progreso showing locations of archaeological units, historic repeat photography, landmarks, and paleoecological column samples.

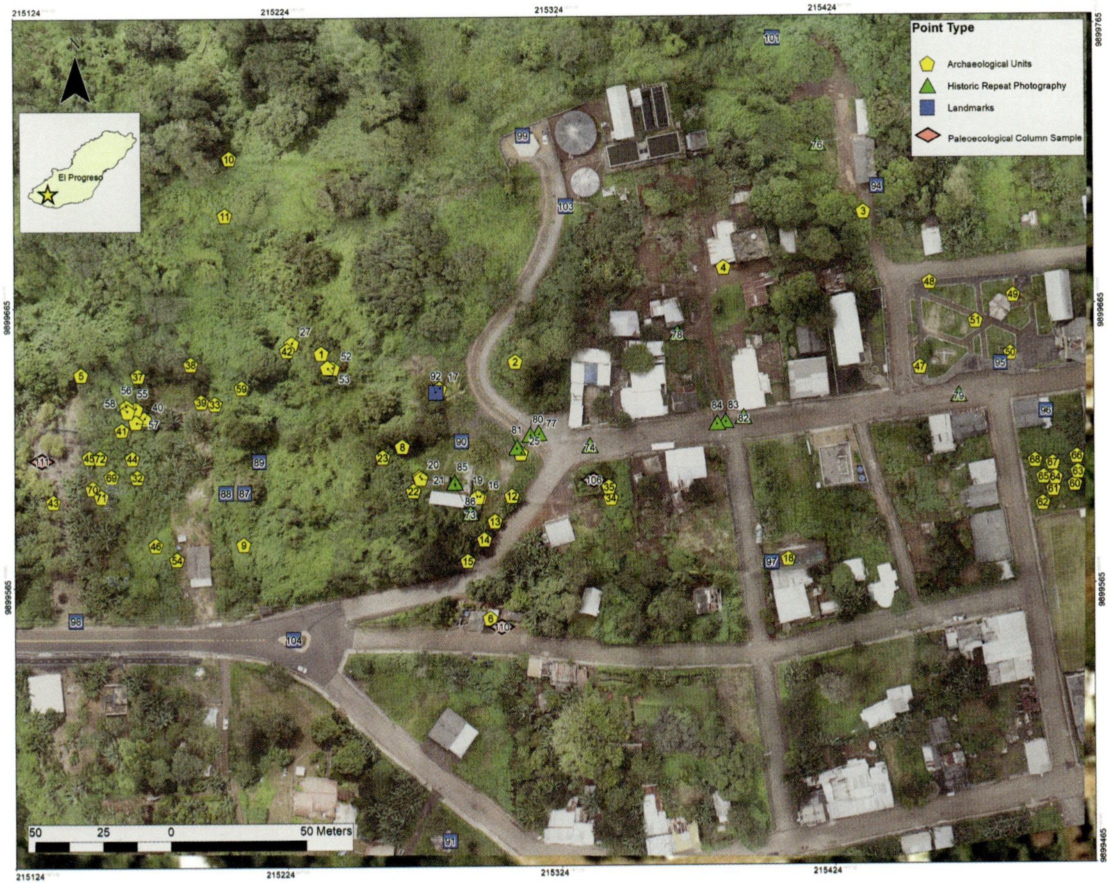

Archaeological Units: Artifact Scatter (1), Big Cistern (2), Camal Unit 10 (3), Camal Unit 11 (4), Canal (5), Carpintero Midden (6), Cobble Pavement (9–11), Cobos Unit 1 (12), Cobos Unit 2 (13), Cobos Unit 3 (14), Cobos Unit 4 (15), Cobos Unit 5 (16), Cobos Unit 6 (17) Cobos Unit 6b (18), Cobos Unit 7 (19), Cobos Unit 7b (20), Cobos Unit 8 (21), Cobos Unit 9 (22), Cobos Unit 12 (23), Datum (25), End of Path (27), Sugar Mill Installation (32), Intact Pipe (33), La Carcel Units 1&2 (34), La Carcel Unit 3 (35), Corners of Large Water Retention Pool (37–40), Little Wall (41), Metal Artifacts (42, 43), Metal Tank (44), Metal Tank with Grates (45), Structure (46), Park Unit 1 (47), Park Unit 2 (48), Park Unit 3 (49), Park Unit 4 (50), Park Unit 5 (51), Paved Path (52, 53), Pavement (54), Corners of Small Retention Pool (55–58), Stone Wall (59), Test Pits 1–8 (60–68), Metal Tubes (69, 70), Tunnel to Smokestack Base (71), Wall (72); **Historic Repeat Photography**: Location 1 (73), Location 2 (74), Location 3 (75), Location 3b (76), Location 4 (77), Location 5 (78), Location 6 (79), Location 7 (80), Location 8 (81), Location 9 (82), Location 10 (83), Location 12 (84), Location 13 (85), Location 14 (86); **Landmarks**: Old Road, Camino Antiguo (87–89), Hacienda House Site (90), Cobos Tomb (91), "El Horno" (92), El Progreso Camal or Abbatoir (94), El Progreso Park (95), HEGI Project Houses (96, 97), Entrance to Sugar Mill area "Larry's Shooter Fire" (98), Mirador Loma del Consuelo (99); Old Road, Camino Antiguo (101); Potable Water Plant (103); Sugar Mill Gears in El Rondel Traffic Circle (104); **Paleoecological Column Samples**: Clm 1 (106), Clm 5 (110), Clm 6 (111). Map Source: Walsh, Stephen J, Joe Eyerman, Carlos F. Mena, Philip H. Page, Justin Dee, Merlin Benner (July 4–9, 2015), Galapagos UAS Data v.10, San Cristobal and Coastal Environments, Galapagos Archipelago of Ecuador; Thermal, Multispectral, and Optical Systems, Data Captured from Multiple UAV Platforms (Fixed-Wing, Quad & Hex Systems), San Cristobal Island, Galapagos Archipelago, Ecuador.

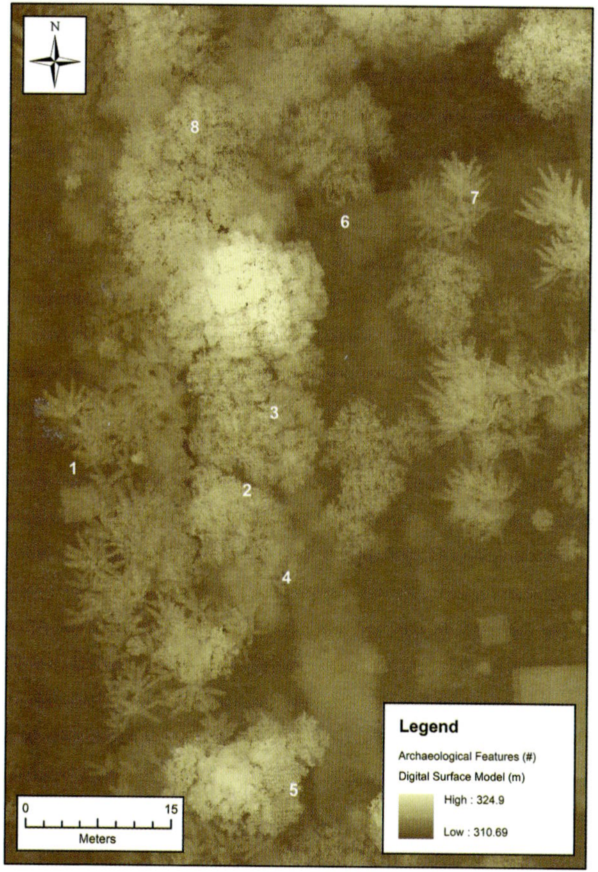

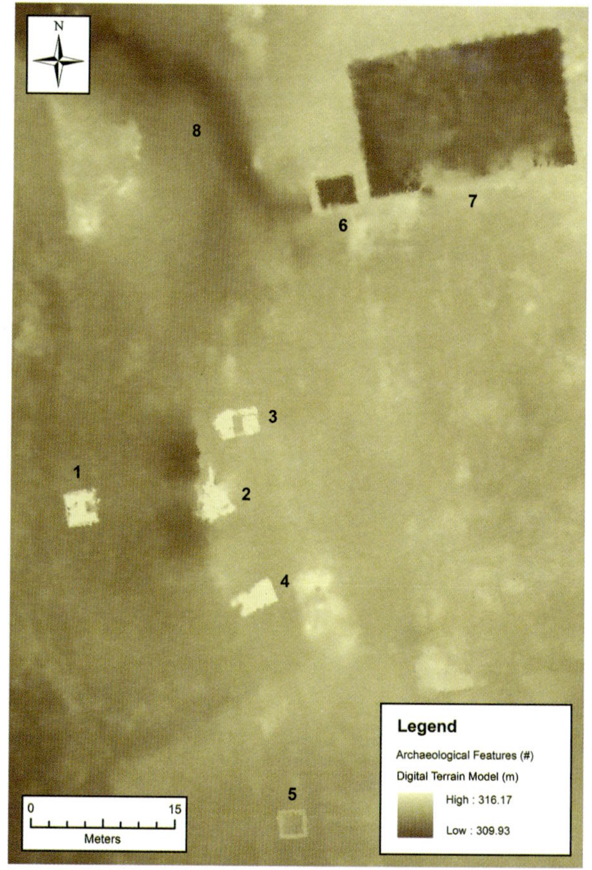

Far left: Plate 2. Georeferenced orthorectified aerial image derived from aerial LiDAR data collected by an unmanned aerial vehicle (UAV) of the central mill area immediately to the west of the hacienda house. Image by Georgia Clyde.

Left: Plate 3. Digital terrain model (DTM) of the central mill area built using LiDAR returns classified as ground and archaeological artifacts, derived from aerial LiDAR data collected by an unmanned aerial vehicle (UAV). Elevation is recorded in meters above sea level (MASL). 1. Smoke Stack Base; 2. Furnace Area; 3. Boiler; 4. Boiler; 5. Small Stone-lined Water Cistern; 6. Deep Stone-lined Water Cistern; 7; Large Stone-lined Water Cistern; 8. Earthen Canal. Image by Georgia Clyde.

Plate 4. Repeat Photo #1, Location #13. Historic Photo: 1888 "Plantation Chatham Island 1888" National Archives and Records Administration (Albatross Expedition), NARA-22-FA-88. Repeat Photo #2231–2241, View ENE, GPS: S00°54.451' W089°33.484' Elevation: 337 m.

Plate 5. Repeat Photo #3b, Location #11. Historic Photo: 1888 "Fertile Plateau of Chatham Island Seen Looking East from the Cobos Hacienda" National Archives and Records Administration (Albatross Expedition), NARA-22-FA-89. Repeat Photo #2190–2200, View ENE, GPS: S00°54.378' W089°33.436', Elevation: 342 m.

Plate 6. Repeat Photo #7, Location #2. Historic Photo: "no information" 1905 Rollo Beck G71, California Academy of Sciences, Rollo and Ida Beck Collection, Box 40 MSS036, Repeat Photo #1707–1713, View E, GPS: S00°54.436' W089°33.472'.

Plate 7. Repeat Photo #8, Location #1. Historic Photo: 1919 "The Peon's Quarters" Ralph Stock (1921) *Cruise of the Dreamship*. London: W. Heinemann, National Geographic Image Collection/Granger, Repeat Photo #2139–2148, View E, GPS: S00°54.438' W089°33.475'.

Plate 8. Repeat Photo #9, Location #5. Historic Photo: 1940s? "Viaje Presidencial de San Cristóbal (Chatham)," Photo attributed to C. I. Pazmiño, Digital Files of the Bibliotéca El Progreso, courtesy of Edy Bismark Becera Hernández, Repeat Photo #1730–1739, View ESE, GPS: S00°54.433' W089°33.430'.

Plate 9. Repeat Photo #10, Location #4. Historic Photo: 1933/34, "Chatham Island Village of El Progreso," C. McLean Fraser (1943) *General Account of the Scientific Work of the Velero III in the Eastern Pacific, 1931–1941. Part II Geographic and Biological Associations*. Allen Hancock Pacific Expeditions 1(2). University of Southern California Publications, Los Angeles. USC Libraries. Allen Hancock Foundation Collection, Repeat Photo #1721–1726, View E, GPS: S00°54.434' W089°33.433'.

Plate 10. Repeat Photo #13, Location #12. Historic Photo: 1888 "Natives, Chatham Island 1888" National Archives and Records Administration (Albatross Expedition), NARA-22-FA-93. Repeat Photo #2214–2222, View ESE, GPS: S00°54.445' W089°33.487'.

Plate 11. Repeat Photo #14, Location #12. Historic Photo: 1888 "Natives, Chatham Island 1888" National Archives and Records Administration (Albatross Expedition), NARA-22-FA-92. Repeat Photo #2223–2230, View ESE, GPS: S00°54.445' W089°33.487'.

Figure 9. El Progreso sugar factory. This picture was likely taken in 1927, more than 20 years after Cobos' death, when a significant superstructure was still standing. (View SW-NE). Hoff (1985). Photo courtesy Nette Næss.

The El Progreso *ingenio* required enormous volumes of water to drive the state-of-the-art machinery that converted field cane into a consumable and exportable sugar product.[13] Water was transported from interior highland sources to the processing plant in an elaborate and lengthy system of canals, aqueducts, conduits, and pipes.[14] Today, a large inground water retention basin, measuring approximately 16 × 9 m and constructed of thick volcanic rock and marine sand mortar walls more than 1 m wide and 2–3 m high, dominates the northern end of the mill area. An opening at the bottom of the deepest corner of the basin transferred water into a smaller and deeper adjoining water reservoir with a metal tube drain imbedded in its west wall (Figure 10). Directly across and downslope from the adjoined basins is the stub of a volcanic stone and mortar structure with a tunnel opening connected to the base of the smokestack (Figure 11). Conceivably, water pressure increased with flow from the large shallower basin into the constricted deeper basin, toward boilers heated above the structure with the underground tunnel. At least one other smaller basin is found on the property, all serving in a larger water retention system interconnected via earthen canals and metal tubing preserved today in

Figure 10. Stone water cisterns.

Figure 11. Stone structure with opening to underground tunnel connected to the smokestack.

Figure 12. Stone structure with arched furnace opening leading to a deep combustion chamber.

the immediate area. A large and slightly depressed basin located directly north of the sugar processing area can become inundated during heavy rains, and it may have been connected to the processing area by an earthen canal, portions of which are still visible under the thick vegetation.[15]

The *ingenio* required a massive and constant supply of combustible fuel to process the field cane immediately after it was harvested, since the onset of decomposition was rapid. The largest surviving feature on the property today is an approximately 3.5 $m^2 \times 4$ m high structure that may have been part of the tall central building that dominated the historic *ingenio* (Figure 12). This structure may have been vented by the chimney in the historic photo. Constructed of finished volcanic rock and mortar, it has an arched opening on its western side which leads into a deep pit that likely functioned as the combustion chamber. Whether or not the chimney also vented the steam boilers is unknown; however, large quantities of heated water were required to generate the necessary steam for driving the equipment used in processing sugar.

Two of the three large boilers mentioned by Martínez in his 1906 visit, which supplied the steam that powered the machinery, are conspicuous today on the *ingenio* property (Figure 13). The boilers, one with the upper half of its midsection removed, lie to the south of the water basins in close

Figure 13. Intact boiler.

proximity to each other. Nearby, a low structure built of volcanic stone and mortar with furnace door openings on one end (Figure 14) likely served to support the boilers which at one time were mounted on top. The multiple drilled openings at either end of the boilers, combined with supporting structures housing combustion chambers, variously referred to as dutch ovens or fire boxes, suggest that the necessary steam required to operate the press engine and other processing equipment was supplied by a system of horizontal return tubular (HRT) fire tube boilers.[16]

Regularly in use today, this system included the standard and most widely distributed American boilers during the early nineteenth century, which were popular for their affordability, compactness, low overhead space requirement, and high capacity (Gebhardt 1928: 121). Also referred to as smoke tube multitubular boilers, they were preferred for their greater water capacity, and importantly the underlying Dutch ovens were large enough to completely combust bagasse fuel (Deerr 1911: 406–408). Lying next to a smaller retention basin is the cast iron door to one of the combustion chambers (Figure 14), which bears the embossed manufacturer's identification, "The Walsh & Weidner Boiler Co. Chattanooga Tenn USA." This company, formed in 1889 to produce pressure vessels and both fire and water tube boilers, was subsequently consolidated as Hedges, Walsh, Weidner Co. in 1928 (Pare 2010).

Figure 14. *Top*: Boiler support structure with combustion chamber. *Bottom*: Fire box door embossed with "Walsh & Weidner Boiler Co. Chattanooga Tenn USA."

Nothing survives today of the engine that operated the machinery; however, the preserved cane presses suggest some possibilities. These are found today in various locations, including the large spur wheel and pinion that decorate the entrance to El Ceibo, and the wheels and press rollers in the middle of the roundabout of El Rondel (Figure 15). An intact three roller mill is currently on display at the entrance to a hotel in Puerto Baquerizo Moreno, which is operated by a descendant of Manuel Cobos (Figure 15). The mill, a MacDonald design with a "solid" or "closed headstock" type (Deerr 1911: 173), displays the embossed manufacturer's identification, "Geo. L. Squier MFG Co., Manufacturer Buffalo N.Y. USA" on the non-geared side. Modern processing plants at the time usually had a combination of at least three triple roller mills driven by one engine (Deerr 1911: 170).

The George L. Squier Manufacturing Company, originally founded in 1857,[17] specialized in plantation machinery for the tropical market. The configuration of the surviving triple roller mill appears to be most similar to their "Louisiana No. 2," a 9,000 lb machine built for steam or hydraulic power. It was rated for the daily production of between 8,000 and 10,000 gallons of juice and a capacity of 60 to 100 acres of cane per harvest (Squier 1870: 31; 1879: 64). The mill, bolted to a wooden bed, was designed for use with a medium-duty horizontal engine producing between 15 and 25 hp. The Louisiana No. 2 could be purchased in a set, which the company packaged together with their medium-sized "Hercules" engine, graded pans used in a "Jamaica Train," grates, oven doors, a bronze pump, centrifuge, and a saccharometer or sugar hydrometer with glass, for $3,200 (Squier 1879: 123).

Immediately after its extraction the juice was clarified or defecated for further concentration of sugar through separation. Juice, heated to below boiling temperature, could be transferred to large settling tanks with filtering materials and added lime in order to remove unwanted solids. It could also be directly decanted into rectangular eliminators where it was heated to a boil and further clarified through separating light impurities by manually skimming froth, and removing heavy impurities that settled on the bottom (Deerr 1911: 241). The juice was eventually evaporated to a syrup; the George L. Squier Co. illustrate their Deep American Evaporator, consisting of a long iron pan with built-in defector compartment (Squier 1879: 76, 79). The pan was constructed to rest on top of a heated surface that required a low, narrow arch-roofed structure housing an

Figure 15. *Top*: Spur wheel and pinion decorating the entrance to El Ceibo. *Middle*: Press wheels and rollers decorating the middle of the roundabout at El Rondel. *Bottom*: George L. Squier cane press displayed at the entrance to a hotel in Puerto Baquerizo Moreno.

interior fire chamber accessed by a small ash pit door on one end and a chimney structure at its opposite end. It is possible that one such structure, composed of volcanic stone, bricks, and mortar, and lying in close proximity to one of the existing boilers (Figure 16), served this function.

Hacienda El Progreso took full advantage of modern evaporation techniques by utilizing a triple effect vacuum evaporator, which Mann described in his 1907 visit (Mann 1909: 29). Although no material vestiges of the elaborate equipment exist today on the landscape, Beck supplied a photograph of the device, possibly as early as 1905 (Figure 17). The disconnected pipe to the steam source suggests that the evaporator may have not been functioning at the time the image was taken. It is likely that the valuable machine was eventually dismantled and shipped to a mainland plantation. On display in the yard of a contemporary residence across the road from the mill is a metal plate embossed with the manufacturer title "McOnie Harvey & Co. ??60 Glasgow." The company, established in 1870 and incorporated in 1892 as McOnie Harvey & Co., Ltd., specialized in machinery for sugar plantations, especially the triple effect vacuum evaporator, for which it held the patent (Anonymous 1897: 135–136).[18]

Centrifuges were used as a final step after evaporation for concentrating sugar crystals in the remaining massecuite. Introduced in 1849, centrifuges revolutionized sugar processing by replacing the labor intensive curing method employing drainage through clay, and later metal, molds (Moreno Fraginals 1976: 117). A large round metal container with basal flanges and through-holes for bolting into place is used today as a flower planter in the courtyard of El Ceibo. It appears to be the bronze basket container of a No. 3 Centrifuge marketed along with the Louisiana No. 2 cane press (Squier 1879: 83).[19] Processed sugar was next moved in large wheeled boxes, or *gavetas*, to a drying area where it was stored in drying racks for a number of months (Tezanos Toral 2015: 85–86). Cane alcohol or *aguardiente* byproduct was produced via adventitious fermentation of the exhausted molasses, often in a nearby liquor loft with large fermenting vats and either direct-fired or steam-operated stills (Deerr 1911: 515).

Some *ingenios* transported sugar products to the purging house in small gauge railway systems; however, no surviving remnants are found in the mill area today. Firsthand accounts describe a small gauge Decauville railway used to transport harvested field cane to the mill, consisting of 50 cars pulled by oxen along 7 km of track (Bognoly and Espinoza 1905: 166; Martínez 1915: 47). Traction-powered rail systems designed for

Figure 16. Arched support structure for a possible evaporator with combustion chamber opening, and boiler in background.

Figure 17. A triple effect vacuum evaporator (California Academy of Sciences G71.5, Rollo and Ida Beck Collection, Box 40, MSS.036). Photo taken by Rollo Beck in 1905 or 1906 (Slevin 1931).

Figure 18. A Decauville rail system wheel in the collection of a hotel in Puerto Baquerizo Moreno.

Figure 19. View of the historic wharf with Decauville rail system and platform truck with iron basket, 1919. Photo, Ralph Stock (1921) *The Cruise of the Dream Ship*. London: William Heinemann, facing page 118: "watering with Dingy and Sand Flies." National Geographic Image Collection/Granger.

transporting harvested intact cane stalks from the fields employed long platform trucks with 8 wheels on 20 inch gauge tracks of 14 lb rails with fixed sleepers. Each cart with iron baskets held up to 1,300 lbs and was pulled by oxen tethered with long yolks on each side of the track (Decauville 1884). Two surviving wheels, embossed with "Decauville Aine Petit-Bourg IS80," and found near the *ingenio* site, are today in the collection of the hotel with the cane mill (Figure 18).[20] Carts hitched to two or three teams of oxen transported export products to the large warehouse in Puerto Chico for shipment to the mainland. The historic pier, whose rock foundation survives today at los Marinos on the southeastern corner of Wreck Bay, appears to have been traversed by the small gauge tracks of a Decauville railway to load and unload ships bound for the mainland (Figure 19).

The Zooarchaeological Legacy of Ranching and Animal Exploitation at Hacienda El Progreso

Exotic animal domesticates had been stocked on San Cristóbal Island since at least the earliest years of Ecuador's Republican Period; by the 1840s, and after the demise of Asilo de la Paz, cattle had become commonplace. Beef and cowhide were exploited for local consumption and to provision whalers; however, upon his return, Cobos began to corral the island's feral herds, estimated by some to number in the thousands.[21] Shortly after his death, the amount of improved and unimproved pasturage on the highland's western slopes was estimated to range from 1,000 ha and 2,500 ha to more than 4,000 ha (Bognoly and Espinosa 1905: 166; Mann 1909: 29). In addition to sugar, molasses, and cane alcohol, the hacienda exported salted fish, animal hides, and oils to the mainland. Introduced, domesticated products were shipped to the coast from the interior highlands, and various endemic, wild products were transported to the interior for local consumption.

The preserved animal bone samples recovered from the excavations around El Progreso help to reveal the hacienda's catholic consumption. Archaeofaunal specimens routinely made up the bulk of materials in number and in weight (Table 1). Although also collected from the Camal and Cárcel units, only the 23,025 (109,451 g) specimens from secure deposits in the Carpintero midden are considered for the interpretations considered here.[22] Despite the relatively small portion of excavated midden, the

Table 1. List of archaeofaunas recovered in excavations from El Progreso

El Progreso Archaeofaunas	Common Name	Freq.	Wt. (g)	MNI
Indeterminate	Indeterminate	37	26	
Anthozoa	Coral	1	3	
Eucidaris	Pencil Urchin	24	7	
Chiton goodalli	Giant Chiton	1,290	2,627	
Chiton sulcatus	Sculptured Chiton	166	196	
Mollusca	Mollusk	36	44	
Gastropoda	Gastropod	2	2	
Tornatellides cf. *chathamensis*	Tree Snail*	1	-	
Subulina cf. *octona*	Miniature Awlsnail*	1	-	
Bulla	Bubble Snail	2	5	
Cancellaria gemmulata	Sculptured Nutmeg	1	1	
Cantharus sanguinolentus	Sanguine Cantharus	1	2	
Cerithium	Cerith	4	27	
Cerithium gallapaginis	Galapagos Cerith	2	-	
Conus	Cone	3	43	
Conus nux	Nut Cone*	2	19	
Cypraea albuginosa	White-spotted Cowrie*	3	16	
Hipponix grayanus	Beaded Hoofshell*	1	-	
Plicopurpura	Purpura	3	66	
Plicopurpura columellaris	Small-Mouthed Purpura*	1	39	
Plicopurpura patula	Wide-Mouthed Purpura	3	106	
Patellogastropoda	Limpet	1	-	
Fissurellidae	Key Hole Limpets	6	2	
Fissurella	Key Hole Limpet*	1	-	
Lottiidae	Limpet	9	1	

El Progreso Archaeofaunas	Common Name	Freq.	Wt. (g)	MNI
Bivalvia	Bivalves	2	-	
Arcidae	Ark Clams	1	-	
Papyridea aspersa	Clam*	1	10	
Lucinidae	Lucine	1	-	
Ctena	Galapagos Lucine	1	1	
Ostreidae	Oysters	1	1	
Pinctada	Panamanian Pearl Oyster	1	17	
Pectinidae	Scallops	1	4	
Osteichthyes	Bony Fish	1,682	1,089	
Serranidae	Groupers	1,174	2,481	40
Reptilia	Reptiles	9	-	
Cheloniidae	Marine Turtles	774	5,074	3
Iguanidae	Iguanas	20	13	2
Aves	Birds	37	8	
Gallus	Chicken	40	43	4
Mammalia	Mammals	18,105	65,538	
Carnivora	Carnivores	36	13	
Canis	Dog	66	179	3
Felis	Cat	15	11	2
Artiodactyla	Two-Toed Ungulates	244	2,322	
Sus	Pig	65	363	3
Bovidae	Bovids	1	2	
Capra	Goat	234	1,606	6
Bos	Cattle	1,369	36,838	27
Equus	Horse	2	35	1
Rodentia	Rodent	3	-	
Rattus	Rat	5	-	2
Leporidae	Rabbit	1	-	1

Note: * = taxa not identified in Carpintero midden; Freq. = total numbers for entire sample; Wt. = cumulative weight in grams (≤1 g) per individual specimen; MNI = Minimum Number of Individuals calculated only for vertebrates from the Carpintero midden.

identified fraction included large quantities of mammal bone fragments, most probably derived from cattle, as well as abundant specimens of large sea bass or grouper.

Exotic Domesticated Animals

The heavily fragmented sample of large mammal and bovine specimens implicates the importance of cattle production for both local and export consumption. Locally consumed beef, and hides and dried charqui produced for export, were regularly obtained by dedicated hunters harvesting wild highland cattle and from tamed animals penned in improved pastures. The limited available evidence preserved in the historic midden suggests that these were relatively small-statured animals probably of the nondescript Criollo landrace, an often tan-colored bovine with short, fine hair, barrel-shaped body, and upswept horns (Rouse 1977).[23] Today, the feral Galapagueño cattle are considered one of seven biotypes of the Ecuador Criollo (Porter et al. 2016).

The analyzed sample of recovered cattle bones suggests that almost half the animals were slaughtered under the age of 4–5 years, at least four of which were butchered possibly as young as 2 years, and others at more advanced ages.[24] The sample in the Carpintero midden appears to represent the deposition of waste products from locally butchered animals. Elements from the entire cow skeleton were recovered in the Carpintero midden, albeit in different proportions; however, their differential survivorship was neither mediated by relative bone mineral density, nor correlated with nutritional value.[25] Historic records emphasize the production of skins for export, thin sheets of sun-dried charqui, and rendered fats, all forms of production that took place on other islands as well. Visible surface modification as butchery scars and chop marks made with machetes and axes suggest removal of the head and jaws, disarticulation of limb elements, skinning of hides from lower limbs, and extensive destruction of the vertebrae and ribs during preparation for consumption.

Goat specimens are relatively common in the midden sample and indicate the presence of at least two different breeds, a smaller and possibly native Criollo, and a larger, likely Nubian variety.[26] The 28 native breeds of Criollo, including the Criollo del Ecuador, and the partially feral Galápagos goat, both descended from stock brought to the western hemisphere on Columbus's second voyage in 1493, and introduced via early colonial Panamá. Easily adapting to hot climates and in global demand for cross-breeding, the Nubian was stocked as shipboard food on P&O steamers and disseminated at ports-of-call around the globe during the second part of the nineteenth century, often for milk and meat (Ginja et al. 2017; Porter et al. 2016: 356–357). Various dismemberment marks were identified on elements at joints, in addition to modifications on the hip and rib shafts. At least one individual was slaughtered under the age of 1 year, with the remainder under 5 years.[27]

Limited evidence attests to the breeding and slaughtering of pigs at Hacienda El Progreso; identified specimens recovered from the midden indicate the processing of relatively young swine within a maximum of 3 years, and generally between 1.5 and 2 years.[28] Pigs share a similar early Columbian pedigree with cows and goats. Genetic studies of Criollo pigs indicate a complex phylogeny and an interestingly closer genetic relationship between Ecuadorian and Cuban Criollos, which may indicate a genetic bottleneck after Cuba became an important distribution center during Conquest (Burgos-Paz et al. 2013; Revidatti et al. 2014). Historic sources mention that horses and mules were abundant, but the midden yielded only two, extremely worn, equine teeth. The introduction of horses to Galápagos may have included the Colombian Paso Fino from an early Caribbean insular source via Panamá, or perhaps the less selectively bred Criollo de Vaquería from the eastern llanos (Jiménez et al. 2012).

Household animals are also identified in the Carpintero deposits. Historic sources frequently attest to the scourge of dogs on all the inhabited islands. Although recovered dog specimens were relatively common, the sample may represent the disposal of two puppies and one beagle-sized dog, which appears to have a healed fracture of its front foot. Similarly, two domestic adult cats older than 1 year were deposited in the midden.[29] A few hind limb elements, likely from introduced rats, were tentatively identified, as was the mandible of a rabbit. Limb elements, many concentrated in the lower levels of one excavation unit of the Carpintero midden, attest to the presence of domesticated chicken.[30]

Endemic Wild Animals

Preserved specimens of larger marine fish are abundant in the historic midden. The sample includes roughly equal numbers of heavily fragmented and unidentifiable marine fish, and identifiable sea bass, most likely larger groupers.[31] Cranial and postcranial elements were recovered in similar proportions, suggesting that entire fish were transported to the hacienda from the coast. Galápagos Grouper or Bacalao (*Mycteroperca olfax*) was the primary targeted species, especially after the establishment of commercial fin-fishing in 1945, with groupers and sea bass making up the bulk of island catch. However, larger apex-level groupers have declined dramatically in importance as local fisheries increasingly fish down the marine food web, with herbivorous mullets (Mugilidae) and coastal pelagic fish now assuming greater economic importance (Schiller et al. 2015).

Written sources indicate that the historic hacienda was actively pursuing turtles, especially for oil. A large volume of midden bone is composed of marine turtle specimens that may have been brought to the interior highlands for local consumption.[32] Biomolecular analyses of recovered turtle elements confirm the identification of Green Turtle (*Chelonia mydas*).[33] Although the relevance of this limited genetic study to populations existing more than 100 years ago is unknown, the sampled individuals possess haplotypes similar to offshore feeding aggregates (Royle 2017). Green Turtles spend their juvenile phases in the open ocean, and after attaining sufficient size during upward of 10 years, they return to coastal waters and become predominantly herbivorous (Arthur et al. 2008; Bolten 2003). Recent genetic analyses identify the CmP97.1 haplotype as belonging to orphans closely related to foraging (i.e., nonbreeding) haplotypes in the West Pacific and suggesting trans-Pacific associations (Chaves et al. 2017).

It is interesting that no Galápagos tortoises were identified in the midden, although reports from around 1860 suggest that island tortoise populations were already exterminated on Floreana and severely diminished on San Cristóbal. However, the studied sample is but a tiny portion of the industrial-scale midden, and tortoise may have been shipped live or their remains disposed of elsewhere. Similar speculations can be raised by the absence of other exploited animals, especially sea lions, which would have been procured on the beach, and subsequently processed locally, with preserved bone specimens deposited in unsampled coastal contexts. A few iguanas were identified, mainly from vertebral specimens. The San Cristóbal Marine Iguana (*Amblyrhynchus cristatus mertensi*) inhabits rocky coastlines, intertidal areas, and marine waters as deep as 20 m.

A variety of common, shallow-water, and principally intertidal marine gastropods, bivalves, and sea urchins were recovered in smaller amounts from the midden. Two species of larger chitons are relatively abundant, particularly the Giant Galápagos Chiton that congregates in groups of up to a dozen individuals in the narrow cracks between lava in the intertidal zone. Ubiquitous chiton shells are today scattered around the town site as villagers regularly access either coastline to harvest prized *canchalagua*.[34] Intertidal habitats were exploited for other edible marine invertebrates like oysters, scallops, clams, lobster and crabs, many of which cling to rocks exposed at low tide. Other, nontidal species like Bubble Snails and Coral may be present in the midden as the result of accidental by-catch.

The Archaeobotanical Legacy of Sugar Production, Farming, and Ranching at Hacienda El Progreso

Analyses of preserved archaeobotanical specimens recovered from archaeological and landscape contexts provide an opportunity to study the ecological impact of human colonization on the environment of San Cristóbal Island. Archaeological data can be used to examine the early relationship between native highland vegetation, human colonizers, and the exotic crops they introduced to support industrial-scale agriculture and farming in order to continue their dietary traditions from the Ecuadorian mainland. Historical records can be combined with an analysis of preserved botanical macroremains, wood charcoal, and phytoliths to explore crop introduction and local diet; verify historic fuelwood collection by identifying woody taxa utilized during the plantation's operational years; and understand possible deforestation and the impact of colonization on the native vegetation of San Cristóbal Island, at least since the 1860s.

Exotic Crop Plants

Charred seeds recovered from the archaeological midden provide evidence for the early importation of common mainland crops and fruit trees. A total of 80 charred seeds and seed fragments were extracted from soil samples associated with plantation contexts (AD 1870–1920). Charred kernels of maize (*Zea mays*) were the most prominent identified taxon (n = 35), demonstrating its existence on San Cristóbal Island at least since the 1870s. Maize was likely one of the first crops introduced into the archipelago by its early colonizers, who numbered highland and coastal farmers among them. Maize is predominantly listed in the historical sources; during his 1906 visit, Martínez (1915: 30) noted that it was grown year-round in large quantity on exceptional parcels of land.

Charred coffee (*Coffea arabica*) seeds were also preserved in the midden. Martínez (1915: 30) emphasizes the grand-scale cultivation of excellent quality coffee on a considerable expanse of land in 1906. Coffee production was an important activity at Hacienda El Progreso during the 1890s, and by the time of his visit, 172 cuadras (ca. 170 ha) of land were dedicated to its cultivation (Compañía Guía del Ecuador 1909). Other charred seeds identified in the midden assemblage include lentils (*Lens culinaris*), a popular edible legume in the Andean region. Nutritious, easily transported, and quick to prepare, lentils may have been imported for local consumption. Seeds of the Common Guava (*Psidium guajava*) are identified in the midden. First introduced to Galápagos during early colonization, it rapidly became invasive throughout the entire archipelago and is now one of the major threats to native vegetation (Guézou et al. 2010). Today, dense patches of guava trees dominate large areas of the San Cristóbal highlands.

Soil phytoliths recovered from the historic midden layer also reveal the presence of maize and sugarcane (*Saccharum officinarum*).[35] The introduced plant taxa identified in layers above the concentrated midden lens present challenges for the interpretation of phytolith concentrations, as they may represent periods of abandonment or later disturbance through plowing or construction. These contexts record: an increase in saddle and rondel phytoliths, which are possible indicators of sugarcane; phytoliths of banana (*Musa*);[36] globular echinate phytoliths associated with palms;[37] and a re-emergence of globular phytoliths that may represent abandonment periods and invasive vegetation resilience. Mill contexts record a permanent presence of grass species, represented by large concentrations of short bodies. It is possible that analyzed layers with ash and charcoal were deposited in the midden as fuel waste from regularly cleaned boiler combustion chambers used to generate steam for powering the *ingenio* machinery. The presence of grass phytoliths from sugarcane leaves and stems in the midden ash would accord with the observations made by visitors to the hacienda that bagasse was burned as fuel after crushing (Mann 1909: 31).

Woody Taxa

Clearance of vegetation for pasturage began in the 1860s when Cobos and Monroy stationed workers on the island to prepare the land for agricultural activities. The wood charcoal assemblage in the midden records the presence of both native and introduced taxa on San Cristóbal Island since that decade.[38] Fragments of native Matazarno (*Piscidia carthagenensis*), a tree or shrub that can grow up to 15 m in height throughout the archipelago's arid lowlands and moist uplands, dominate the assemblage. Highly valued for its hard inner wood and preservative properties, it is likely that Matazarno was used from the early years of Galápagos colonization as the main source of timber for construction materials.[39]

Endemic Galápagos Guava or Guayabillo (*Psidium galapageium* var. *howellii* on San Cristóbal Island, McMullen 1999: 83) is a small tree or shrub that grows up to 8 m tall. Its wood has been, and is still, used for firewood, fencing, and small tool making. Although still preferred for firewood, native stands of Galápagos guava have been replaced by the related exotic invasive Guava (*Psidium guajava*). Endemic Tree Scalesia or Lechoso (*Scalesia pedunculata*) reaches heights up to 20 m and possesses soft wood and gummy sap (McMullen 1999: 45). Found in small stands around the humid San Cristóbal highlands, it forms extensive forests on Santa Cruz and Isabela. Tree Scalesia could also have formed sizable forests in the highlands of San Cristóbal prior to the establishment of permanent human populations. Its trunk and branches can be used as firewood, and it is possible that native Scalesia forests were reduced on the southern end of the island after vegetation clearance for agriculture, pasturage, and sugar production beginning in the 1860s.

With a worldwide distribution, bamboo is one of the most commonly used plants in tropical environments, where the widely distributed Neotropical Bamboo (*Guadua angustifolia*) is used for construction, scaffolding, fencing, small-scale irrigation systems, and furniture manufacture. Likely introduced into Galápagos during colonization as a cheap and reliable construction material, bamboos adapt rapidly and grow quickly with little maintenance.[40] A few medium fragments (<10 mm diameter) of pine (*Pinus* sp.) and oak (*Quercus* sp.) were identified; neither is native to Galápagos or northern South America. Martínez (1915: 32) reported two specimens on San Cristóbal in 1906 that were apparently ornamental trees. Oak is also mentioned in other early descriptions of the plantation

(Mann 1909; Webster 1904). A plausible explanation for its incorporation as burned fragments in the hacienda midden may include burning imported wood in the form of furniture or wooden boxes commonly used to transport the many fragile household items of porcelain and crystal, liquor bottles, and tableware also abundant in the Carpintero midden.

Two other native trees are possibly present in the wood charcoal assemblage. Galápagos Croton or Chala (*Croton scouleri*) is a small 2–6 m tall tree or shrub endemic to the islands, mainly at lower elevations, and is used as firewood. Incense Tree or Palo Santo (*Bursera graveolens*) is a 3–12 m tall tree or shrub native to the arid lowlands in the islands. With an aromatic resin that produces a distinguishable odor when broken or burned, Palo Santo is regularly used for incense in churches throughout Latin America. Also distributed along the South American coast, it is used as a natural insect repellent (Jørgensen et al. 1999). Both species could have been used mainly for firewood, small tool making, gardening, fencing, and as incense.

Impact on Native Vegetation

In addition to providing direct evidence of introduced plants, the analyzed soil phytolith assemblages from Hacienda El Progreso's midden and mill offer indirect indications of the island's landscape prior to the arrival of permanent human occupants. The phytolith assemblage recovered from below the midden is associated with mixed vegetation; whereas, concentrations in the historic lens suggest the possible increase of grassland vegetation and the introduction of exotic grasses possibly for expanding pastures.[41] The phytolith record from the mill suggests a permanent presence of grass species represented by large concentrations of short bodies. It is likely that grass phytoliths in the analyzed ash and charcoal layers derive from sugarcane leaves and stems burned as bagasse fuel for the steam-powered machinery of El Progreso. Modern vegetation is represented by high concentrations of *Musa*, palm, and globular phytoliths in the superficial layers. These plants were growing on the surface when the soil samples were collected.

Paleoecological column samples were extracted from four 60 × 60 cm test pits placed in different contexts of the agricultural zone: village, abandoned field, forest, and agricultural field (Astudillo 2018c). Altogether, 52 samples were collected in 5 cm levels excavated to a depth of 60 cm.

A Tree Cover index (D/P) was calculated for each context to measure vegetational structure in absolute and relative terms by graphing the ratio of certain forest phytoliths to grass phytoliths (Strömberg 2009).[42] The D/P indices vary between 5.54 and 0.21 in contexts associated with the plantation and illustrate uniform changes in local vegetation composition (Figure 20). The results show higher counts of globular phytoliths and other arboreal-derived phytoliths in earlier contexts, and large concentrations of grasses after human arrival on the island. Globular phytoliths are formed in leaves and stems of native taxa that early accounts (California Academy of Sciences 1907; Compañía Guía del Ecuador 1909; Mann 1909; Stewart 1911) describe as important components of local Galápagos Island landscapes.[43] After human arrival, grasses became a primary component of the local landscape around the early settlements. Arboreal vegetation dominated the landscape prior to human arrival and was partially replaced by grasses, likely during the second half of the nineteenth century. Land clearance for intensive agriculture is the most likely reason for the observed alteration in vegetation cover.

These activities would corroborate the sudden increase of grass-derived phytoliths in the analyzed column samples and support paleoecological records that suggest an increase of Poaceae pollen grains during the first decade of the twentieth century (Bush et al. 2014; Collins and Bush 2011; van der Knaap et al. 2012; van Leeuwen et al. 2008; Vargas et al. 2012). The increase in pollen from Poaceae was also observed in a recent soil core from El Junco Lake, 7 km northeast of the study area (Restrepo et al. 2012). This may have been the time when introduced grasses start becoming part of the local landscape, especially *Pennisetum* and *Brachiaria*, which are both recognized in the phytolith record through the presence of the cuneiform bulliform cells, and polylobate and saddle morphotypes. The high numbers of Panicoideae-derived phytoliths in superficial levels of the Scalesia Zone may also be indicating the presence of economically useful taxa such as sugarcane (*Saccharum officinarum*) or bamboo (*Guadua* sp.).

Historic and Modern Landscape Transformations as Seen through Repeat Photography

A small workforce had been permanently stationed on San Cristóbal by Manuel Cobos and José Monroy during the 1860s, in part to improve the

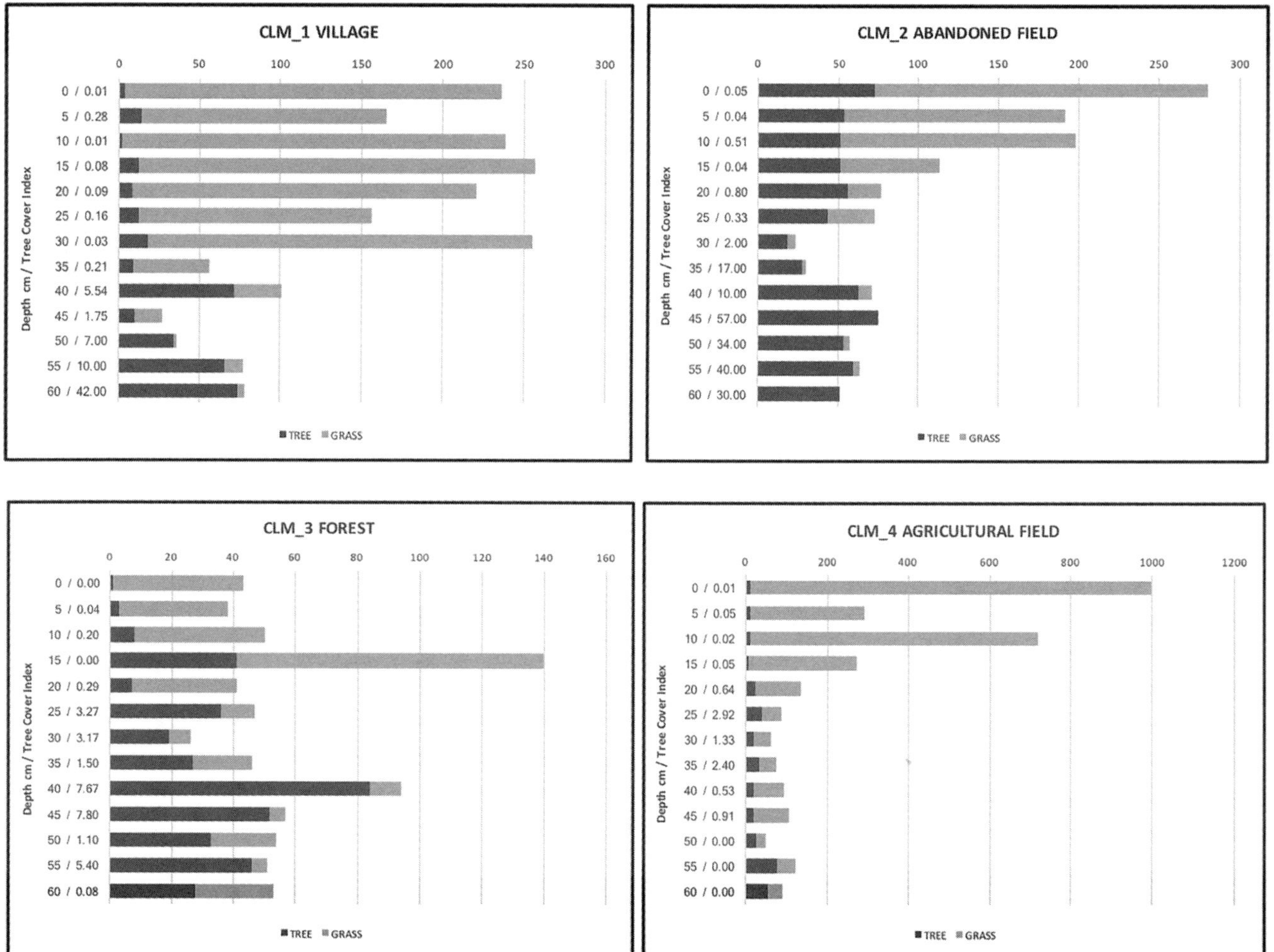

Figure 20. Concentrations of arboreal vs grass phytoliths presented as D/P ratio signals in soil phytoliths from San Cristóbal Island, Galápagos. Note: Uniform change from arboreal to open habitat.

island landscape for cattle pasturage. By 1879, the Hacienda El Progreso was a thriving enterprise, after which Cobos's full-time residency stimulated the commencement of agricultural intensification that led to extensive modification of the island's highland landscape. By the late 1880s, not counting pasturage, at least 117 ha of land was under intensive crop production, including more than 56 ha of rectangular sugarcane fields (Latorre 2002: 15). At its height, various estimates of managed landscape exceeded 1,200 ha, including 700 ha of natural or unimproved pasture, 147 ha of foraging or improved pasture, and 435 ha of cultivated crops (Bognoly and Espinosa 1905: 166). Moreover, sugarcane fields were carefully controlled, supplied with water via a 6–7 km long irrigation system, and constantly weeded in order to maintain high productivity. The enterprise was serviced by a network of roads and trails stretching from the interior highlands to the coast in order to harvest, process, and transport hacienda products.

Shortly after Cobos's death, visitors described productive orchards, gardens, coffee plantations, and productive cane fields, backed in the distance by extensive pastures sloping into the mountains (Martínez 1915: 46–48). Likely exaggerating, Mann (1909: 29) estimated more than 10,000 acres of pasturage extending to the mountain's summit at almost 2,700 feet. More conservative estimates at this time describe approximately 700 ha of pasturage with natural grass, and 150 ha with "Janeiro" grass surrounding El Progreso village (Compañía Guía del Ecuador 1909). Expanding highland pasturage gradually increased deforestation outward from the urban core of El Progreso to the east. Trees, bushes, or trenches were used as fencing since early colonial times in Ecuador.[44] Martinez (1915: 33) expressed concern over the expansion of grasslands. Worried about the impacts of deforestation and the rapid disappearance of natural forests brought about by cultivation and the demand for wood products, he recommended the immediate introduction of fast-growing eucalyptus, pine, and cedar to replace the disappearing forest.[45]

A number of surviving historic photographs taken of Hacienda El Progreso during the late nineteenth and early twentieth century visually depict open pasturage in the distant hills to the east of the historic village. Through repeat photography,[46] the juxtaposition of these historic images with their precisely aligned contemporary counterparts provides dramatic visual reference, both to the extent and degree of vegetation clearance and to the regeneration of the contemporary landscape (Plates 4–11). Today, highland vegetation has colonized the area within the historic footprint of Hacienda El Progreso to form a legacy landscape dominated by exotic taxa introduced from the historic hacienda's earliest beginnings, which continues into the present day.

5

Consumption and Control in the Material Culture of Hacienda El Progreso

During the last quarter of the nineteenth century, Hacienda El Progreso was physically remote from the perceived global centers of "modernity," but the material culture deposited in its midden reflects strong ties to liberal, modern ideas. Latin American elites in this period focused on the modern, which was seen as an idea of universal human progress (manifest in Cobos's case through the name of El Progreso itself), and expressed through imported European goods that represented a break from the Spanish colonial past (Moreno Tejada 2016; Capello 2011).

The years from 1870 to 1930 saw Latin America brought into the global market as exporters of commodities and importers of manufactured goods. Items such as metals from mines and Caribbean sugar had been exported since the colonial period, but the last quarter of the nineteenth century saw unprecedented global market integration of Latin American commodities, as the chaos of the independence period gave way to a free market export boom, moved by steamships and railways (Beckman 2013: ix). Latin American elites gained access to credit and finance for commodity ventures, whereas industrial development did not occur to the same extent as in Europe and North America. The spectacular growth of European markets after the 1850s created huge demand for tropical agricultural products and raw materials, with a combination of capital and skilled labor seen as the key to unlock the vast potential of an untapped landscape throughout the region (Beckman 2013: 4). The Hacienda El Progreso was one small example of such an untapped landscape.

The flip side of the export boom that has received less attention is the massive importation of consumer goods to Latin America in exchange for these commodities. These goods in the late nineteenth century were an important part of bringing modernity to the region, as liberal elites created "modern" homes through their purchases. Modernization in late nineteenth-century Latin America has often been characterized as tied to export production, and secondarily through often failed attempts at industrialization. The importation of consumer goods has been treated as a lamentable side note (Bauer 2001: 1). The late nineteenth-century Latin American elite focus on furnishing their houses and setting their dining tables with products from Europe has been characterized as inauthentic, denying the autonomy of Latin American cultures (Beckman 2013: 45).

The archaeological materials recovered from excavations reflect the wide selection of European and North American goods that defined imported elite culture in the Latin America of the 1890s (Bauer 2001; Orlove 1997). France, in particular, became a fascination for Latin Americans after 1860, and it remained so until the First World War tremendously changed global trade patterns (Andrade 2012; Bunker 2012; Orlove 1997). These liberal "fantasies of progress" were bolstered by the growth of print advertising, pushing foreign imported consumer products as the way forward to a new liberal world order (Beckman 2013).

Cobos's operation probably acquired most of its imported supplies in Guayaquil, where steamships delivered European goods to retailers along the *malecón*, although some may have been acquired through his contacts and travels to California, Baja California, and Panamá. Thus, although the hacienda was a long boat ride from the coast, Cobos, as with all landholding oligarchs of the time, was connected to a global trade in European and North American consumer goods that were used to reinforce a "cosmopolitan" refinement in dining and drinking habits (Gaitán Ammann 2011). Cobos was not unusual in this, as the newly wealthy cacao planter families of the Ecuadorian coast in the 1880s and 1890s sent family members, and a lot of wealth, overseas to Paris, New York, and Madrid, not only to buy fashions or wine, but also to send their children to school and sometimes to acquire a second home in Europe (Chiriboga Vega 2013: 218). Our analysis of the excavated materials from Hacienda El Progreso suggests two distinct aspects to the manufactured goods that were imported: consumption to project a modern image, and technologies to control the landscape and the workers on the hacienda (Meyers and Carlson 2002).

The Contexts

We have divided our analysis of the hacienda's material culture into four separate areas of the site (Table 2). On the west, small-scale area excavation revealed a cobble floor or road bed, and all artifacts (*n* = 57, wt. = 1.5 kg)[1] recovered derive from the soil above this floor. This is such a small sample it is not particularly useful for comparison. Adjacent to Cobos's main house, several 1 × 1 m units in the house gardens revealed a fair number of artifacts to depths not exceeding 50 cm; the lower levels of these contexts appear to date to the Cobos occupation and are included here (*n* = 1,604, wt. = 9.5 kg). In the "workers village," we have merged the artifacts from two excavation areas, one in the middle of the location where worker housing stood in Cobos's time, and the other closer to the area of the hacienda's communal kitchen and outbuildings. In both cases modern materials near the surface were not included in our analysis, but some depth of deposit, again to about 50 cm below surface, revealed materials consistent with the Cobos era, which are included here and grouped as "workers village" materials (*n* = 1,889, wt. = 10 kg). The fourth area lies to the south of the house and consists of the obvious nineteenth-century Carpintero midden deposit, with a dense lens of artifacts. Because this lies south, and downhill, from both the main house and the communal kitchens, it seems likely this was a centralized trash disposal area for the core operation of the hacienda, and we assume the majority of these artifacts came from Cobos's household, but a considerable number may also have come from discards from the communal kitchen and other outbuilding operations. In total we analyzed 7,604 artifacts weighing 80 kg from the midden.

Dating archaeological material deposited in the second half of the nineteenth century has the advantage of containing many items with short date ranges of production, based on the rapidly changing manufacturing techniques of the machine age. This collection is no exception. In the case of all the contexts apart from the midden, stratigraphy was limited to a current surface lens of modern materials, and a distinguishable single occupation under that, probably consisting of mixed materials from throughout the hacienda's occupation. Although important in telling the hacienda's story, these materials cannot be placed into too strict a time frame. The exception is the midden context, where a single lens of capped, artifact-rich material appears to indicate a surface used as a

Table 2. Artifact counts and weights, grouped into modified Parks Canada "activity" categories

Activity	NISP	% NISP	Weight, g	Activity	Weight, g	% Weight	NISP
INGENIO AREA							
construction materials	21	37%	413	machine part unknown	708	43%	9
nails	11	19%	89	construction materials	413	25%	21
machine part unknown	9	16%	708	other fasteners	373	23%	5
glass beverage containers	7	12%	10	nails	89	5%	11
other fasteners	5	9%	373	pharmaceutical	41	2%	1
ceramic tableware	1	2%	1	general storage	28	2%	1
general storage	1	2%	28	unknown	563		6
pharmaceutical	1	2%	41	**TOTAL weight**	**1,652**		
transportation	1	2%	1				
unknown	6		563				
TOTAL NISP	**57**						
WORKER VILLAGE							
glass beverage containers	553	29%	2,334	glass beverage containers	2,334	23%	553
nails	296	16%	1,144	machine part unknown	1,380	13%	1
ceramic tableware	240	13%	538	nails	1,144	11%	296
construction materials	66	3%	1,012	construction materials	1,012	10%	66
glass storage containers	60	3%	227	ceramic tableware	538	5%	240

Activity	NISP	% NISP	Weight, g	Activity	Weight, g	% Weight	NISP
pharmaceutical	51	3%	205	other fasteners	370	4%	8
ceramic cooking/storage	16	1%	124	general storage	265	3%	8
clothing fasteners	10	1%	8	glass storage containers	227	2%	60
unknown	535		2,239	pharmaceutical	205	2%	51
TOTAL NISP	**1,889**			hand tools	139	1%	3
				ceramic cooking/storage	124	1%	16
				unknown	2,239		535
				TOTAL weight	**10,265**		
Big House gardens							
glass beverage containers	548	34%	3,601	glass beverage containers	3,601	37%	548
construction materials	298	19%	1,047	hand tools	1,144	12%	2
nails	188	12%	934	construction materials	1,047	11%	298
ceramic tableware	183	11%	276	nails	934	10%	188
glass storage containers	76	5%	246	general storage	594	6%	28
glass tableware	40	2%	196	pharmaceutical	425	4%	35
pharmaceutical	35	2%	425	ceramic tableware	276	3%	183
lighting	34	2%	28	glass storage containers	246	3%	76
general storage	28	2%	594	glass tableware	196	2%	40
grooming and hygiene	18	1%	54	other fasteners	162	2%	9

(continued)

Table 2—*Continued*

Activity	NISP	% NISP	Weight, g	Activity	Weight, g	% Weight	NISP
ceramic cooking/storage	11		62	machine part unknown	157	2%	1
unknown	110		576	unknown	576	6%	110
TOTAL NISP	**1,604**			**TOTAL weight**	**9,622**		
Big House Midden							
glass beverage containers	1,829	24%	32,703	glass beverage containers	32,703	41%	1829
nails	1,161	15%	6,096	general storage	6,830	8%	253
construction materials	1,005	13%	6,621	construction materials	6,621	8%	1005
ceramic tableware	901	12%	3,898	nails	6,096	8%	1161
general storage	253	3%	6,830	ceramic tableware	3,898	5%	901
glass containers, unspecified	158	2%	680	hand tools	3,411	4%	7
metal containers	158	2%	832	glass storage containers	1,372	2%	74
glass tableware	103	1%	1,200	sweets and indulgences	1,285	2%	59
unknown	1,515		10,396	glass tableware	1,200	1%	103
TOTAL NISP	**7,604**			other fasteners	1,178	1%	42
				unknown	10,396		1515
				TOTAL weight	**80,537**		

Note: Left-hand columns show all activities that represent more than 1% of collection from each site area by NISP. Right column represents same data rearranged to represent all activities making up more than 1% of collection by weight in grams.

trash dump, probably burned several times to inhibit pests and odors, and eventually abandoned, with slow accumulation of garden soils on top of it after the practice of dumping trash there was discontinued. We do not know if the dump was in use for months, or perhaps several years, but it appears to represent a fairly specific, single period of time. Many of the items in the midden deposit have manufacturing date ranges that begin in the 1860s/70s and persist until the First World War or so, and it seems logical to suggest (given time lag from manufacture, through use, to discard), that the material was deposited between 1880 and 1914. With the rapid change in material culture that occurred globally over this period, more accuracy than this is possible. Items in the collection with relatively late starting dates of manufacture represent the earliest date the lens in which they were deposited could have been formed. In the case of the Carpintero midden, key artifacts include a "Nitro Club" 12-gauge shot shell made in 1891–1911, and a Winchester 30-30 cartridge made in 1895 at the earliest. Both of these were recovered from the midden unit 1 level 7. Above this, in unit 1 level 5, a Peters 12-gauge shot shell made after 1887 was recovered.

The nails recovered from the midden are another key indicator of chronology. If we group all the nails recovered from the midden (counting nails and nail fragments) we count 1163 wire nails, 111 wrought nails, and 16 cut nails. The wire nails are by far the most common in all levels of the midden. Cut nails began to replace wrought nails in the 1830s, but in many places in the world they were not popular, as local blacksmiths continued to produce wrought nails. The development of steel wire nails, however, coincided with the increasing use of steamships and trains, and wire nails were thus easily available and cheap quite soon after factories had been set up to produce them. In the United States the massive change in the market from cut to wire nail sales happened in 1880–1890, as wire nails flooded the market as cheaper, effective replacements for cut nails. Their adoption was very regional, depending on market access to particular factories (Adams 2002). In England the market began to change in the 1860s as wire nails from Belgium began to enter the British market. As with the United States, however, it seems the 1880s was the time the British market switched over almost entirely to wire nails (Sjögren 2013). Presumably Cobos purchased his nails from Guayaquil, but the nail trade in South America is poorly understood, and they may have been manufactured in the United States, in Birmingham, England, or perhaps in France or Belgium. In essence, we believe it is safe to state that the midden must have been deposited after 1895, but probably not too many years after that, since items with an initial manufacturing date after 1895 are nonexistent. A date of 1895 or 1900 for this midden suggests that the trash in the midden agrees with Cobos's occupation of the hacienda.

Keep One Hand on the Revolver

One of the most important technological developments in opening new regions to agricultural development in the late nineteenth century was the invention of new, more efficient, small arms. We recovered a total of 40 spent cartridges and shot shells from Cobos-era archaeological contexts: 36 from the main house midden, 4 from the garden surrounding the house, and 4 from the workers village below the house.

In this collection of cartridges we can see the variety of arms that were present. Our recovery of eighteen .44-40 cartridge casings shows the reliance of the hacienda on this caliber as the flexible cartridge of choice. These were used in Winchester repeating rifles, Colt revolvers, and a wide variety of other late nineteenth-century firearms (Houze 2011; Walter 2006). Eight .38 Smith & Wesson casings may be from Cobos's personal handgun. Nine .43 Spanish casings as well as a single 11 × 50R mm Comblain and a single 7 mm Mauser cartridge all reflect the post-1880 Latin American market for European rifles such as Mauser, Mannlicher, and Comblain. Two .44 Henry cartridges are specific to an 1860s weapon, showing occasional use of outdated technology. Two 12-gauge shotgun shells round out the assemblage. Many of Cobos's purchases may have come from the flow of ex-military firearms, slightly out of date, that were sold for the Latin American sporting and self-defense markets (Grant 2007).

No doubt the main use of these weapons was to control livestock pests, such as the abundant feral dog population, and to hunt for "wild" foods such as feral cattle and goats, but they also served a key purpose in maintaining Cobos's control over the human population. Cobos's death at the hands of his workers in 1904, and the subsequent investigation, reveal a number of details about the hacienda's firearms. The workers testified that over the years Cobos had executed five of the workers by shooting them (Webster 1904: 113–114). Cobos's manager, Elias Puertas, stated that prior to the day of the 1904 revolt he found a misplaced .38 revolver which was usually locked in Cobos's office desk. He used this to shoot Cobos. Wounded, Cobos was able to get to a Winchester carbine in another room, and it was this he used to shoot at the workers below his house. When it jammed he returned inside to get another rifle, but the workers had broken in and seized all the weapons (Webster 1904: 115–122). The commission that visited after his death determined that he kept all arms locked in a secure room in the main house, allowing the small police contingent one rifle each when on guard duty. The accountant for the hacienda was in charge of both the accounts of workers' debts to Cobos, and signing in and out the small arms on hand to those who had permission to use them. Guns and account books were both kept in a locked space near Cobos's bedroom (Webster 1904: 51). When the workers eventually escaped on the boat, they took 67 guns with them, leaving 6 guns so the police at the hacienda could still maintain order (Webster 1904: 122–123). There were probably between 75 and 100 guns in Cobos's armory.

Control of the El Progreso workforce was not just through threats of violence. After the workers killed Cobos, one of their first acts was to burn the hacienda's account books (Webster 1904: 139–140), presumably because these held the key to understanding how much debt each worker owed, and were a focus of the workers' rage. In the midden remains we uncovered two glass inkwells, and a stoneware ink bottle impressed with "ADRIAN MAURIN / PARIS" (Figure 21), thus intertwining the role of writing as a technology of control on the hacienda, and as symbol of European material progress (Guerrero 1997; Lyons 2006).

The issue of debt and payment was also materially expressed through the circulation recording of a line beside each worker's name for one day of labor. By the mid-nineteenth century, accounting systems were becoming more modern (Ahumada E. 2010), and this modernization was accompanied by the introduction of both national coinage and widespread systems of tokens specific to particular businesses, for local circulation. Bognoly and Espinosa (1905: 87–102) illustrate the variety of tokens and scrip that Cobos used at Hacienda El Progreso. In our excavations we encountered one of the crude lead tokens, marked "MC," and struck on only one side (Figure 22), that Bognoly and Espinosa claim had a 5 centavo value (Rulau 2000: CLT 1). Interestingly, the five other tokens we encountered were all hard vulcanized rubber (commonly called ebonite or vulcanite), with two examples of orange 20 centavo tokens, and three black 5 centavo tokens (Rulau 2000: CLT 10, CLT 11). These vulcanite examples are all marked

Left: Figure 21. Stoneware ink bottle embossed Adrien Maurin Paris.

Below: Figure 22. Vulcanite (*left*) and lead (*right*) tokens.

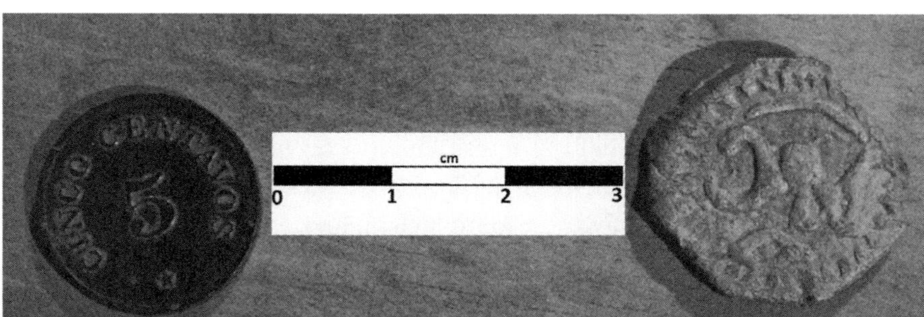

"ANACARSIS MEDINA/CHANDUY" and are illustrated by Bognoly and Espinosa. It would seem these were the main form of coinage circulating on the hacienda during Cobos's tenure, and it is interesting that they are not struck for the hacienda, but instead presumably purchased from Medina's mainland operation in Chanduy. We know little of Anacarsis Medina but assume he may have been an old business contact of Cobos's, as Cobos had operated his business in Chanduy in the years before coming to Galápagos. Bakelite/vulcanite tokens are extremely rare in Ecuador (Rulau 2000), but they were a global phenomenon in the latter half of the nineteenth century. The Peruvian rubber trade and the Chilean nitrate mining sector both used corporate bakelite tokens extensively for payment of workers (Calvo R. 2009), and thus Cobos may have felt he was participating in another very modern administrative practice by importing such tokens to El Progreso.[2]

Another major and archaeologically visible technology of control was barbed-wire fencing. Invented in 1870, barbed wire became a key technology in quickly, and cheaply, being able to fence off huge areas of territory with little labor or expense (Netz 2004; Brantlinger 2018). Presumably imported from the United States, 28 fragments of barbed wire and 29 fencing staples were recovered in our excavations. It was reported that Cobos used barbed wire to enclose many areas of the hacienda (Martínez 1915: 50).

Consuming Modernity

Many of the excavated materials express aspects of consumer behavior at the hacienda, presumably largely by Cobos himself, as it is likely he had direct control over the purchase of most of the imported items. Alcohol consumption was an important part of this consumer behavior, and recovered specimens of alcohol bottles dominate the artifacts in the assemblage (Figure 23). In the late nineteenth century, as today, the types and brands of alcohol consumed by people signaled a lot about social class. In the 1860s Ecuadorian President Gabriel García Moreno helped create a national consciousness of the evils of drunkenness, tied to his generally conservative and Catholic administration. This was applied particularly to the lower class, with legislation restricting sale of alcohol in public, largely as a patriarchal move aimed at curbing consumption by indigenous people (O'Connor 2007; Williams 2005).

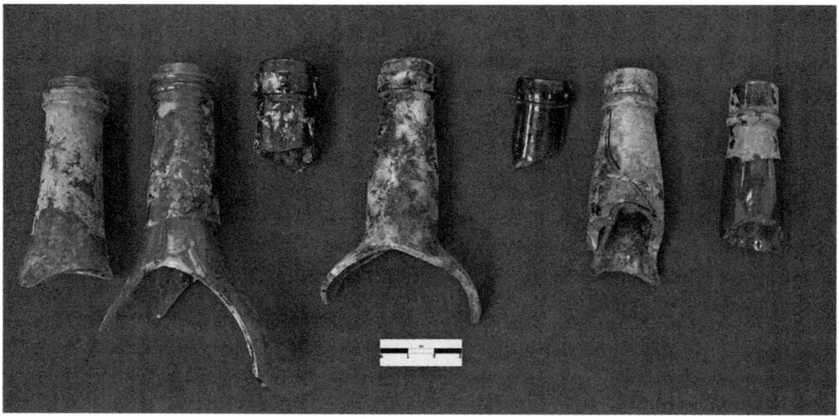

Figure 23. A selection of alcohol bottle finishes.

Race and social class played a strong role in the perceived morality of drinking in late nineteenth-century Latin America, as lower class and indigenous people were assumed to be endangering their ability to labor and raise families through drunkenness on *aguardiente* or moonshine, while the upper classes drank imported and bottled brand-name wines and spirits, a habit tied to "good establishments" and "refined manners" (Carey 2014: 139–141). It is interesting to note that the El Progreso sugar mill was used to produce "alcohol and white rum" (Mann 1909: 29). Whether all of this was shipped off the island, or some was consumed on the hacienda, is not known. In 1907 a visitor described the social tensions that alcohol consumption produced on the hacienda. Martínez (1915: 40) describes that by 1906 problems of drunken behavior were regulated by restricting alcohol sales to workers to Saturday afternoons and Sundays, while maintaining a standard ration of one cup of alcohol per worker per day, each morning before starting work.

The consumption of imported alcohol by Cobos and his staff was noted by many visitors. Martínez (1915: 52) describes a hunting party in 1906 that involved a picnic at El Junco, with bottles of beer and wine submerged in the lake to cool them off. He also pointed out that the *tienda* (company store) at El Progreso had everything the workers might need, including "wine, liquor, and preserves of all types" (Martínez 1915: 39). The most common general category of bottles from the hacienda excavations were wine, beer, or liquor, in a range of colors in greens, ambers, and black. The vast majority were made in turn-paste molds.[3] Many of the larger fragments can be divided into a category of "ale/liquor," as they are typical of shapes that held export beer or hard liquors in this time period, and a separate category of "wine," typically used for wine. We must remain aware, though, that in the late nineteenth century, bottle shapes did not always reflect the standard contents we might expect today. The vast majority of these alcohol bottles would have originally contained an alcoholic product when first sold, but bottlers were somewhat flexible in the shapes and styles used to bottle different drinks.

We have quantified the excavated alcohol bottle fragments from Hacienda El Progreso in two simple ways. First, bottle finishes are used as a proxy for bottles to create an Estimated Vessel Equivalent (EVE). The main house midden produced complete bottles or bottle finishes from one square case or gin bottle,[4] 50 ale or liquor bottles, and 13 wine bottles. From the main house garden we recovered an EVE of 9 ale/liquor and 3 wine bottle finishes, and from the workers village 5 ale/liquor and 2 wine bottle finishes.

A separate vessel equivalent can be calculated from vessel weights.[5] Classifying all glass identified as alcohol-related, and then dividing the total weight of each category by 650 g, allows us to create another way of looking at how many bottles the glass assemblage represents. For the main midden this shows 14 unclassifiable alcohol EVEs, 17 ale/liquor bottle EVEs, and 13 wine bottle EVEs. From Cobos's garden we recovered an EVE of 2 unclassifiable, 2 ale/liquor, and less than 1 of wine. From the workers village there were 2 unclassified alcohol bottle EVEs by weight, 1 ale/liquor, and less than 1 wine. The historical record indicates that a lot of alcohol was consumed at the hacienda—by Cobos, by his staff, and by the workers. In Table 2 it is interesting to note that whether by fragment count or by weight, beverage bottles were the top category of materials recovered from almost all contexts. We think it can be safely concluded from these data that a lot of alcohol was consumed, but actual quantification the amount of alcohol these deposits represent would be very difficult to estimate accurately.

The most common bottle was the ale/liquor bottle, most of which may have contained beer. It is difficult to judge what types of alcohol were purchased in bottles, as most brands were labeled with paper that has subsequently decomposed. There were four bottle fragments with embossed or etched lettering that has survived, each an uncommon type of alcohol we are emphasizing only because these specialized bottles have allowed the information on their contents to survive. The first is a green glass shoulder seal marked "RICHARD & MULLER/NEUFCHATEL" with a Swiss cross in the center of the seal. Ravines (2008: 961) recovered an identical seal from the surface at the (1830–1880) abandoned port town of Islay on the coast of Peru. Ravines lists this as a brandy/cognac bottle, from unreferenced Peruvian import records, while Baker (1951) lists Richard & Muller as exporters of absinthe.

The second, a colorless alcohol bottle base, has acid-etched lettering around the lowest portion of the body, "BARDINET/BORDEAUX." Still in operation, Bardinet has been bottling spirits in France since 1857 and moved to Bordeaux in 1895, giving a maximum age for this bottle. They specialize in rum, but market a wide range of spirits.[6] Finally, two recovered dark green bottle bases were embossed "Dr. J G B SIEGERT & HIJOS" which date from 1872 to 1909 (Fike 1987: 42). These are from bottles of Siegert's Bitters; Johann Siegert was the German-born doctor for Bolívar's troops, who formulated the recipe for this medicinal stomach bitters while living in the town of Angostura (now Ciudad Bolívar) on the Orinoco River in what is now Venezuela. Marketed as "Angostura Bitters," it became a global brand after the company moved production to Trinidad in the 1870s, and remains so today. The recipe is secret, but one or more ingredients appear to have come from indigenous plant remedies known on the Orinoco (Schulz et al. 1980: 64–65). Thus Cobos's drinking habits (and those of many around the world) included not just European products, but also some products created, marketed, and bottled in the Neotropics.[7]

It is clear that beer and/or hard liquor were more popular than wine when looking at imported alcohol bottles on the hacienda. This of course ignores the important matter of bottle reuse. This was a remote place, with the workers living in extreme poverty, so we must assume these bottles were used several times before being discarded. This is shown by a contemporary description of workers' children being baptized on the hacienda. A worker acting as the officiant set up a table in the worker's house, and on it he placed three bottles holding candles, several bottles of *cerveza del país,*[8] and several empty bottles (Bognoly and Espinosa 1905: 104). From this passage we can see that bottles served many purposes in worker households, and would not have been discarded immediately after their imported contents were consumed.

The consumption of imported alcohol in bottles was an elite activity in nineteenth-century Latin America, and yet the bottles from Hacienda El Progreso may indicate signaling of social class somewhat below that of the country's elite. At the Estancia Iraola, the country house of an elite family in 1870s La Plata, Argentina, the trash midden had a bottle glass EVE of 28, with 5 gin/case bottles, 15 wine, and 2 ale/alcohol bottles (Giovannetti and Lema 2007: 87–89). Cobos's assemblage, in comparison, indicates a greater consumption of beer and spirits than wines.

Beyond bottled beverages, it was quite common in nineteenth century Latin America to transport alcohol or wine, along with many other products, in wooden barrels. The Hacienda El Progreso excavations recovered more than 6 kg of barrel strap fragments from the Carpintero midden (Table 2), and a single brass spigot for tapping a barrel. An 1888 photograph of the beach at Wreck Bay (Albatross Expedition, United States National Archive NARA-22-FA-91) shows piles of barrels stacked under the trees, presumably for transport up to the hacienda. We will probably never know the quantity, or variety, of imported goods that came up to the hacienda in barrels.

We also found a large number of fragments from at least one demijohn, a high-capacity bottle with a wide mouth (Figure 24) used to transport wine for decanting into smaller bottles. A demijohn was recovered from the 1770 British naval sloop HMS *Swift*, which was wrecked when blown off course from the Falklands to Argentina (Elkin et al. 2006); in the 1870s midden at the Estancia Iraola in La Plata, Argentina, fragments of a 10 liter capacity demijohn were recovered (Giovannetti and Lema 2007: 87–89). It is thus clear that South American elites were used to the idea of purchasing wine in demijohns, which were usually shipped in a woven basketry cover and/or straw packing and a wooden crate.

A Very Modern Table

Another common category of recovered glassware includes condiment bottles in a variety of types, and none with labels. Two items stand out. The first consists of a series of tall, thin, colorless bottles with very skinny necks (Figure 25). Although unmarked, these shapes are typical of "salad oil" or "olive oil," and published marked examples originated in France during the late nineteenth century, with this bottle shape given in

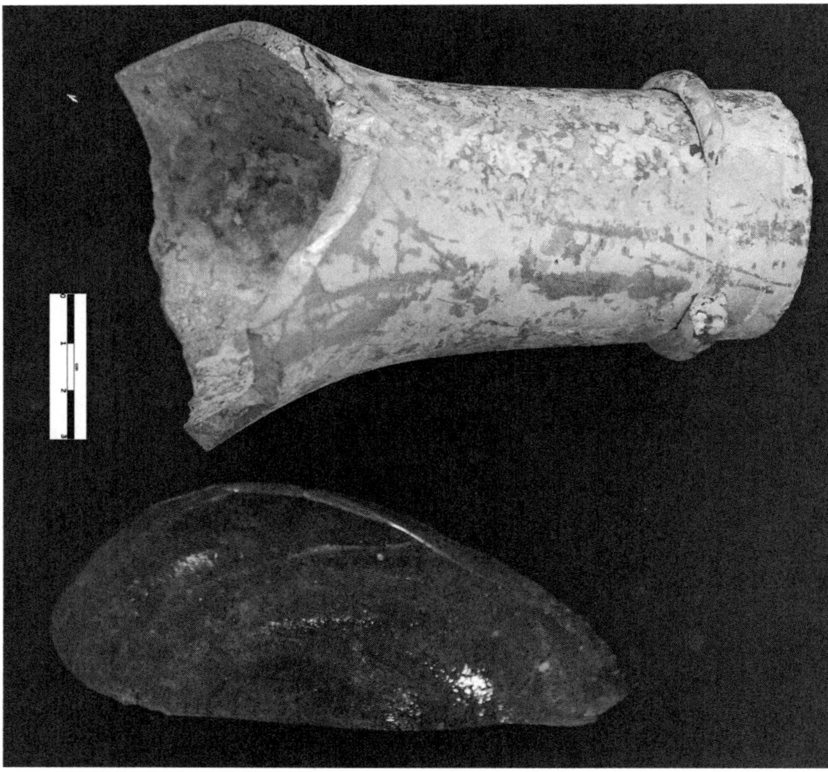

Figure 24. Fragments of a colorless glass demijohn.

contemporary catalogues as a "Bordeaux Oil" shape (Zumwalt 1980: 114, 126). The other major component includes bright green "barrel mustard" bottle fragments; barrel-shaped bottles with horizontal ribs, and "wide patent" rims. As with the oil bottles, these barrel-shaped mustard bottles were frequently imported from Bordeaux in France, as French mustard was considered standard for nineteenth-century consumers throughout the Americas (Switzer 1974: 49; Zumwalt 1980: 448).[9]

The recovered ceramics show a similar affinity for European products. In hacienda excavations of similar date throughout Latin America, a considerable proportion of the recovered ceramics are usually locally made coarse earthenwares. In excavations of late nineteenth century henequen hacienda worker housing at Hacienda Tabi, more than 90% of the ceramics were unglazed, locally made, coarse earthenwares (Meyers 2012: 105–115).

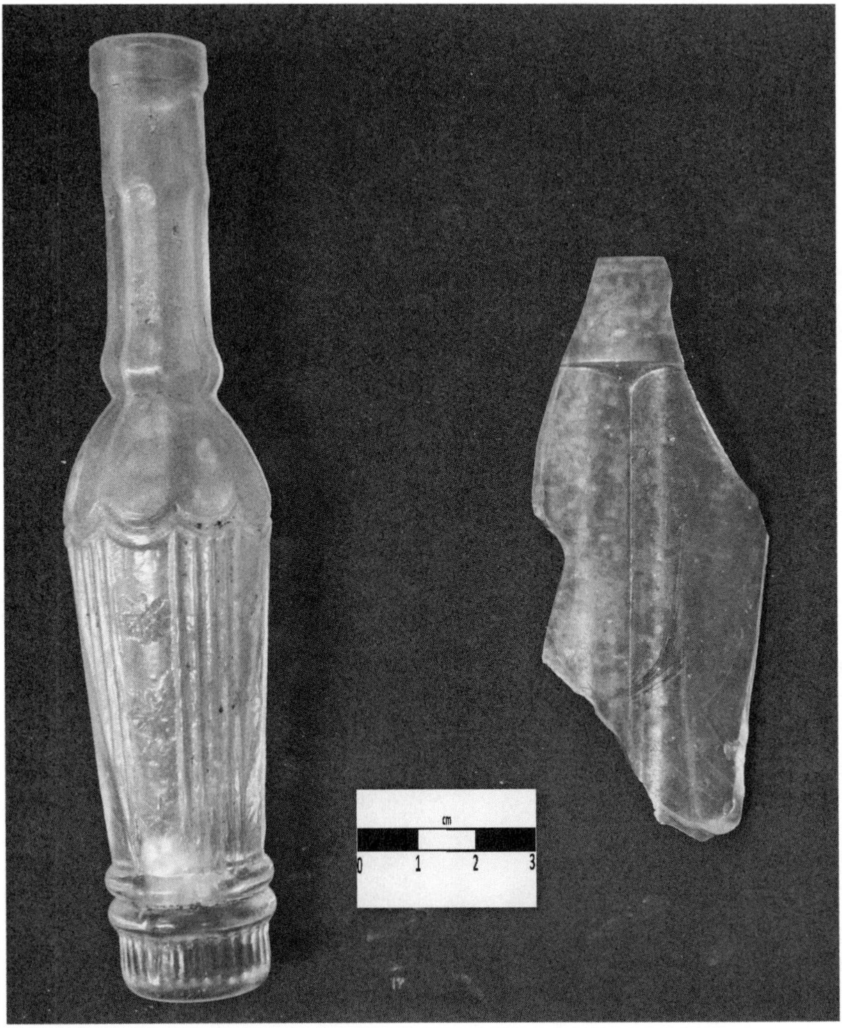

Figure 25. Colorless condiment bottle and a fragment of a tumbler.

The situation was entirely different at Hacienda El Progreso, with coarse earthenwares never making up more than 5% of the recovered ceramics from any given context. We could take this as a sign of greater wealth at Hacienda El Progreso, but instead it is probably more a matter of access. Most mainland Latin American hacienda worker communities had strong ties to local indigenous villages, and to regional market towns, where cheap coarse earthenwares could be purchased. It seems likely that the lack of coarse earthenwares at Hacienda El Progreso was the result

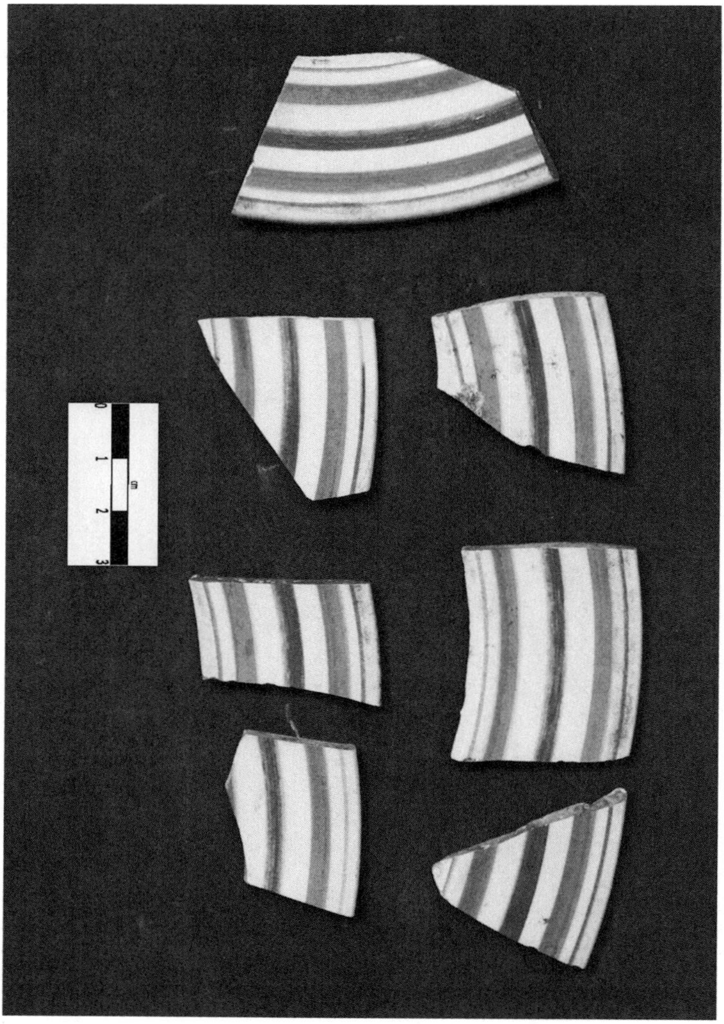

Figure 26. Banded white earthenware tableware fragments.

of the remoteness of the hacienda. Workers were almost entirely cut off from the mainland, and Cobos controlled all imports. Thus the ceramics recovered from excavations were almost entirely from Europe. In this sense "modernity" is not just about wealth, or emulating European styles. It is also about divorcing indigenous peoples from local community ties, and introducing cheap industrial products that can replace locally made artisanal wares.

Published comparative work on late nineteenth-century domestic ceramics in South America is still rare. The refined white earthenwares (RWEs) from Hacienda El Progreso were dominated by industrial lined and banded decoration (Figure 26). By sherd count from all our excavation contexts, plain RWE was most common (*n* = 869 sherds), most of which must represent undecorated fragments of decorated vessels. Sherds with banded decoration (*n* = 200) were the most common of RWE decorations, followed by "Gaudy Dutch" hand painting (*n* = 38), sponge/stamped decoration (*n* = 23), transfer prints (*n* = 20), and finally a teaware set in a variety of monochrome pastels (*n* = 14). This emphasis on banded decoration is typical of the latest styles in Latin American tastes after 1880. Brooks and Rodriguez (2012: 82) briefly examined some post-1860 excavated collections in Venezuela and noted the dominance of French banded wares, particularly from the Vieillard factory.

In our excavations 15 refined white earthenware fragments were recovered with legible makers' marks that could be sourced and dated (Table 3). Five were British, made in the Staffordshire potting district, including three from J&G Meakin, one from Johnson Brothers, and one from Adams. Eight were from France, including three from J. Vieillard & Cie, in Bordeaux, two from Gien, on the Loire River, two from Sarreguemines in Lorraine, and one from Barluet & Cie./Creil & Montereau, from Creil in the north of France. Finally two of the marks were Belgian, both from the factory of Boch Frères, in Saint Vaast-La Louvrière.

A collection of RWE from an elite house midden in the center of Barcelona, Venezuela, dated to the first half of the nineteenth century, was dominated by a set of "Gaudy Dutch" polychrome tablewares, which appear to have been the main tablewares for the house. Shell-edge plates and willow-pattern transfer prints were also common. It appears that this collection of imported ceramics was entirely British, once again demonstrating that prior to 1860, British producers were unrivaled in the Latin American elite ceramic market (Brooks and Rodriguez 2012). Becher's (2011: 36) excavations of a municipal trash midden in Bahía Blanca, Argentina, recovered 10 British ceramic makers' marks on RWE, and 4 from Petrus Regout of Maastricht in Holland.

Excavation of an 1870s midden at the Quinta de Bolívar elite mansion in Bogotá, Colombia, revealed 10% plain porcelain and 17% plain whiteware, with 73% coarse earthenwares (Gaitán Ammann 2011: 154 n.4). There were no decorated RWE sherds, showing strong ties to then-current

Table 3. Refined white earthenware ceramic makers' marks from Hacienda El Progreso excavations

Provenience	Artifact #	Transcription of text on mark	Manufacturer	Date range	Country of manufacture	References
main midden	3615	ROYAL IRONSTONE CHINA/ JOHNSON BROs	Johnson Bros.	1883–1913	England	Godden 1964: 355, mark# 2176.
main midden	4049	IRO . . .	W & T Adams	1890–1914	England	Praetzellis et al. 1983, mark# 10; Godden 1964: 22, mark# 35
main midden	4213	. . . EAKI . . . /HANLEY . . . / ENGLAND	J & G Meakin	1870+	England	Praetzellis et al. 1983, mark# 184; Godden 1964: 427, mark# 2601
main midden	4214	. . . CHINA /. . . . N	J & G Meakin	1870+	England	Praetzellis et al. 1983, mark# 184; Godden 1964: 427, mark# 2601
main midden	3616	. . . NSTONE C . . .	J & G Meakin	1870+	England	Praetzellis et al. 1983, mark# 184; Godden 1964: 427, mark# 2601
main midden	9999	OPAQUE DE SARREGUEMINES	Sarreguemines	1864–1895	France	Kowalsky and Kowalsky 1999: 640
main midden	4017	J. VIEILLARD & CO/ BORDEAUX	J. Vieillard et Cie.	1845–1895	France	Kowalsky and Kowalsky 1999: 641
main midden	4435	BORDEAUX	J. Vieillard et Cie.	1845–1895	France	Kowalsky and Kowalsky 1999: 641
main midden	3613	B et Cie / Deposé / Creil et Montereau	Creil et Montereau/ Barluet et Cie.	1876–1884	France	infofaience.com
main midden	3614	FAIENCE . . . DE GIEN / / DIPLOME D'H . . .	Gien	1875+	France	infofaience.com
workers village	4375	PORCELAINE OPAQUE DE GIEN	Gien	1886–1938	France	infofaience.com
workers village	4376	. . . RREGUEMINES	Sarreguemines	1864–1895	France	Kowalsky and Kowalsky 1999: 640
surface find	4370	BO . . .	Boch Frères	1887–1910	Belgium	Kowalsky and Kowalsky 1999: 634
surface find	4369	B / F / Made in Belgium	Boch Frères	1891–1910	Belgium	Kowalsky and Kowalsky 1999: 634

ceramic styles in the United States, where colorful ceramics were seen as destined for the working class. All four RWE makers' marks recovered in the midden were from Meakin, demonstrating that in 1870s Bogotá the British still dominated the elite ceramics market (Gaitán Ammann 2011: 155–156). In excavations of an 1860s urban middle-class domestic midden in São Paolo, Brazil, all the marked RWE ceramics (10 in total) were British, except for one Dutch example from Petrus Regout. The vast majority of the RWE was decorated in either flow blue or willow transfer prints (Arauja and Carvalho 1993: 83–91).

It would thus appear that Cobos's mix of English and French ceramics was different from most excavated samples of nineteenth-century South American domestic contexts. Cobos was fascinated with France, and in fact he sent his son away to school there. Rather than a personal foible, however, his fascination with France was shared by many in the Latin American elite during the Belle Epoque, and this had caused patterns in consumer goods entering South America to shift, from an almost total emphasis on British products, to one open to imports from a variety of European producers, and most emphatically France.

Sewing

Throughout the Cobos era at El Progreso there were reports that the imbalance between male and female workers created huge social problems. It is clear, though, that many of the workers formed families, and Cobos himself had children. Direct archaeological evidence of these women and children on the hacienda is scarce, but we can propose that the mending and maintenance of clothing may have been the work of some or all of the women on the hacienda (Beaudry 2006).

A rectangular colorless patent bottle embossed "SPERM/SEWING MACHINE/OIL" suggests either the presence of a sewing machine or the use of this oil for other fine mechanical objects such as clocks. The global scale of industrialization is epitomized by this bottle fragment. The sperm whale had been an important factor in increasing global interest in Galápagos, as the archipelago had been used as a whaling stopover throughout the first half of the nineteenth century (Dolin 2007). By Cobos's time this trade was greatly reduced as a result of both the overharvest of sperm whales and, after the 1850s, the replacement of whale oil with petroleum for many applications. Sperm whale oil, however, remained the first choice for fine machine oil in the later nineteenth century. Other examples of this bottle type have been recovered from an African-American laundress's house in New York (Matthews and Manfra McGovern 2018: 46); from Fort Bowie in Arizona (Herskovitz 1978); from the First Nations village in Bella Bella, British Columbia, in a late nineteenth-century household midden (Lynch 2015: 70); and from the wreck of the *Aarhus*, which was en route from New York City to Sydney, Australia in 1894.[10]

The maintenance of clothing is also represented in the collection of recovered buttons. From the Carpintero midden and the house garden, these include 15 bone, 12 mother of pearl, and 13 white Prosser china buttons.[11] Although the oil bottle cannot exclusively be related to sewing machines, and the bottle and buttons do not prove the existence of female sewing labor at the El Progreso hacienda, the remoteness of this place and the need to clothe a large number of people make the existence of at least part-time seamstress labor a clear possibility.[12]

Children Playing

Many people at Hacienda El Progreso, including Cobos himself, had children. Archaeologically, as is typical for late nineteenth- to early twentieth-century assemblages, doll parts dominate the recognizable toys recovered. We recovered one porcelain doll head with black hair and eyes and red lips, along with two legs with brown heeled boots, and one plain doll's arm (Figure 27), all from the main midden. These would have been sewn into a stuffed cloth body, and were probably made in Germany, as that country dominated the global production of ceramic doll parts from 1880 to 1914 (Feister 2009: 110). The brown heeled boot was a common feature of doll legs from the 1880s onward (Feister 2009: 112). We recovered two marbles, a plain doll leg, and one 35 mm diameter plain toy plate for a toy tea set from the worker's village. This seems too small a sample to posit any distribution of toys across the hacienda. Usually associated with instilling ideas of Victorian femininity in girls, these dolls represented miniature adult women, rather than babies (Baxter 2005: 41–46; Feister 2009). They were not very expensive, although it seems likely to surmise that the hacienda administration may not have gone out of its way to import dolls to the island for the workers' children; it is quite plausible that these doll parts came from Cobos's daughter's own doll.

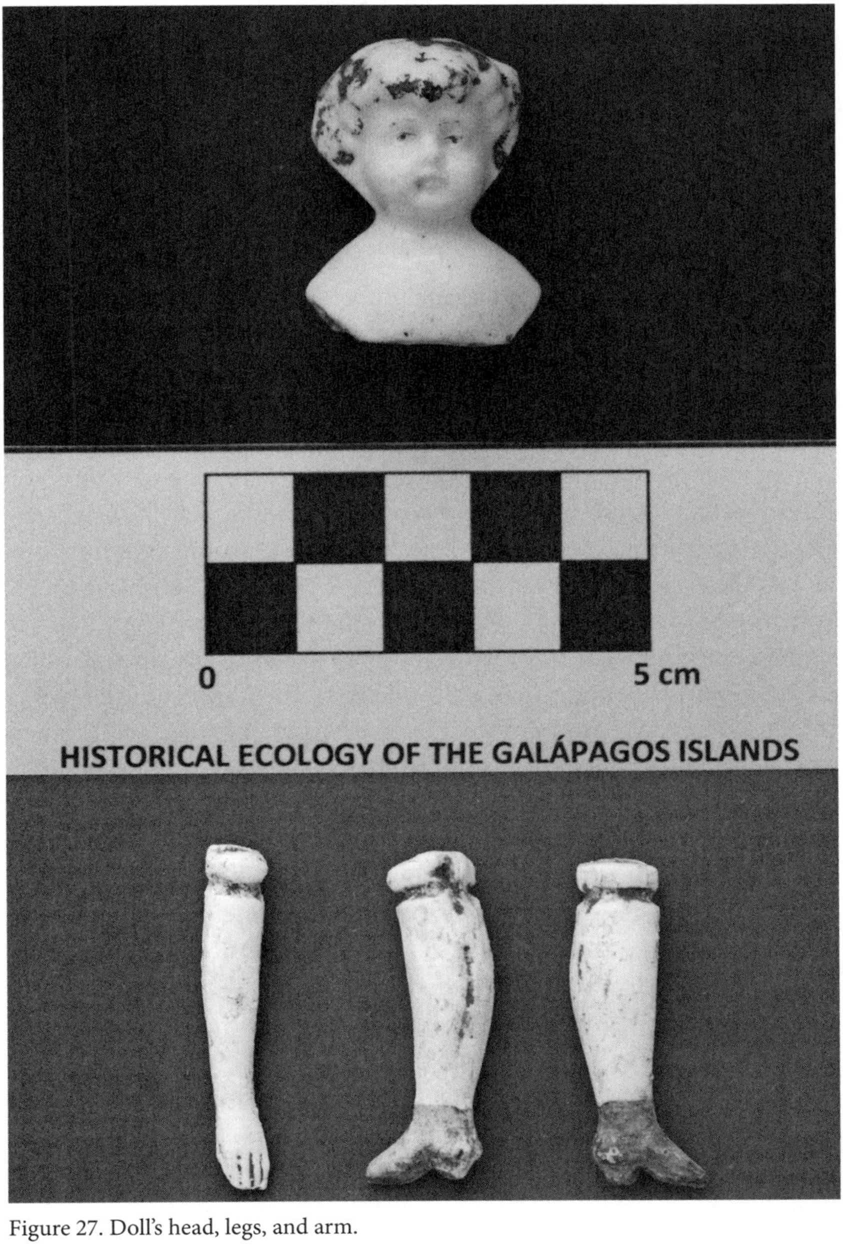

Figure 27. Doll's head, legs, and arm.

Clean and Healthy

The creation of tooth-cleaning products paralleled other areas of late nineteenth-century consumer culture in South America. We recovered a fragment of a toothpaste pot lid matching a "Cherry Toothpaste/Crown Perfumery" lid from London (Dale 1977: 406–407). The sale of recognized toothpaste brands, replacing the purchase by elites of tooth powders that were formulated by individual pharmacies, began in the 1850s, and toothpaste was at first sold in small ceramic pots. They were popularized and manufactured in much larger numbers after 1880, and faded after the First World War as metallic tubes from large toothpaste manufacturers replaced them (Dale 1977: 18). Crown Perfumery in London operated throughout this period, so the lid could date to any time from 1860 to 1918, although likely 1880–1914. Along with evidence for the use of imported tooth powder, we recovered two toothbrushes, one with the word "Superieure," suggesting that at least one of the toothbrushes was made in France. The subject of dental hygiene, and its relationship both to France and to ideas of the modern, is covered well by Gaitán Ammann (2005) based on his recovery of a toothbrush from the midden of the Quinta de Bolívar house in Bogotá. A French toothbrush was clearly a requirement for elite Latin Americans in the second half of the nineteenth century, as evidenced by the recovery of toothbrushes made in France from a number of elite household middens (Schávelzon 1994: 64).

Two examples of "Murray and Lanman's Florida Water" bottles were recovered in the Hacienda El Progreso excavations. This product was made by the Lanman and Kemp Company in New York City and was bottled by 1857. It is probably the most popular brand of "toilet water" of all time, and it has been popular in Latin America since the nineteenth century (Sullivan 1994). It still sells well in the Andean countries but is now used more as an essential product in shamanic curing ceremonies (Foutiou 2012) than as a men's cologne.

Hacienda El Progreso was remote from the mainland and the largest colony in the Galápagos during its operation. Despite this, there was no resident doctor or nurse to our knowledge, and the only mention of health care was the existence of a "hospital" room inside Cobos's house. We have very little information on who took care of the sick or injured, and what was available to them. The late nineteenth century was a time when patent medicines thrived as modern cure-alls, and we recovered

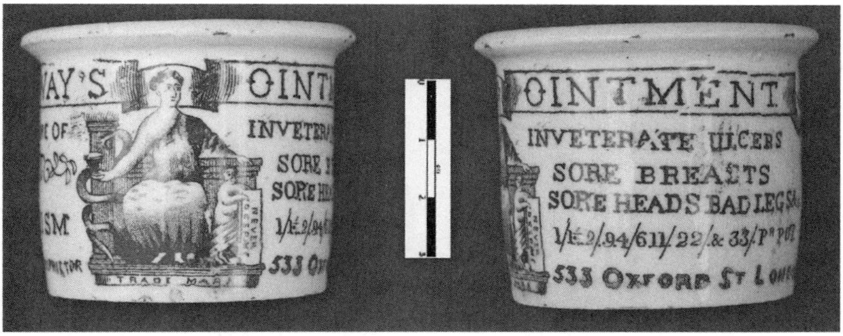

Figure 28. Holloway gout ointment jar.

several dozen fragments of patent medicine bottles, some with embossed lettering that could be identified. One bottle marked "GLYCEROPHOSPHATE ROBIN/GRANULE" is from the Dr. Maurice Robin Laboratories in Bourges, France. They specialized in peptones and phosphates, which were marketed as curatives for nervous and wasting diseases.[13] A bottle of "Dr. H.F. Peery's Dead Shot Vermifuge" was a deworming agent, made by A & B Sands in New York City (Fike 1987: 176). "Dr. D. Jayne's Expectorant," from Philadelphia, was intended to clear the lungs and chest and improve breathing. Fragments from a "Bristol's Pills" bottle, which was also from New York, provide further evidence of Murray and Lanman's drug and toiletry market share in South America. Bristol's Pills were intended to help digestion, preventing headaches, flatulence, and constipation (Fike 1987: 201). Finally, the recovery of a complete Holloway Gout ointment jar (Figure 28) is evidence of the reach of Thomas Holloway, a patent medicine tycoon from London (Holloway 2016).[14] The global reach of this ointment is evident in the recovery of similar Holloway ointment pots from late nineteenth-century contexts in Australia (Nayton 2011: 168), New Zealand (Fraser 2002: 68), and the United States (Linn 2008).

The Chimú Bottle

Very little archaeology has been carried out in the Galápagos Islands prior to our work in El Progreso, and most previous work has focused largely on attempts to prove, or discredit, the idea that people may have visited, or inhabited, the islands prior to Tomás de Berlanga's 1535 "discovery" of the archipelago. Coarse earthenware sherds with molded decoration,

Figure 29. A coarse earthenware bottle fragment from north coast Peru.

recovered from several levels of the Carpintero midden, provide important evidence in this debate. These sherds, when refitted, are from one bottle (Figure 29). This specimen has a dimpled background very similar to the "San Juan Molded" sherds previously recovered by Heyerdahl and Skjölsvold (1956) in the Galápagos Islands. They found these sherds at James Bay on Santiago Island, and these, along with other identifiable North Coast Peruvian prehispanic ceramic styles, were used as evidence of pre-European human habitation of Galápagos.

The partial vessel we recovered appears related to North Coast Peru Chimú styles, and can be tied to the ongoing colonial Peruvian production of coarse earthenwares in Chimú-Inka-Colonial styles (Carlos Elera, pers. comm. 2018; VanValkenburgh et al. 2017). The recovery of this large portion of a North Coast Peruvian vessel in a securely dated late nineteenth-century midden deposit provides further key evidence for the assertion by Anderson et al. (2016: 181) that the Chimú-Sican materials used by Heyerdahl and Skjölsvold as their strongest evidence of prehispanic human arrival in Galápagos are much more likely to have been curiosities brought over to the islands at any time from the 1500s up to the beginning of the twentieth century.

A Remote Place

The Hacienda El Progreso was one of the remotest places on earth from the perspective of someone from London or Paris in the 1890s, yet its material culture shows an amazing range of items from Europe and North America. Modernity was very much on order if, like Cobos, one had cash and a connection to the steamship trade in 1890s South America. His consumer choices (and we assume they mostly were his personal choices, and that managers and workers probably had little say in what was shipped to the island) tended toward European, and particularly French, liquor, while his small arms were largely American. Patent medicines, too, came mainly from New York, although with some European contributions. For setting the table, ceramics were British and French. We do not know how he made choices as a consumer, but we presume most of these choices were made from items available in the shops of the port of Guayaquil, and they may in many ways reflect the consumer habits of the planter class of coastal Ecuador at the time. Cobos's commitment to bringing all these products to his Galápagos highland operation, however, demonstrates an amazing commitment to the world of consumer goods in such a remote location, and it offers a strong contrast to our image of Cobos as brutal master over a village of workers living in crushing poverty with few material comforts.

6

Galápagos, San Cristóbal, El Progreso, and *Colonos* in a Changing World

After its first recognized encounter with humans early in the sixteenth century, Galápagos was introduced into a larger global sphere of human influence. El Progreso, far from being an isolated village in the middle of the Pacific Ocean, was to eventually become the central node that connected humans and the islands with a global system of trade and biological, economic, cultural, and social participation. It is a connection that continues to influence the archipelago.

Humans have been shaping and managing their environments for millennia. Galápagos is no exception, as humans have been actively influencing local island landscapes for nearly 500 years. Nevertheless, over the past decades, Galápagos has come to be considered the pristine remnant of a natural world that existed outside the later and disturbing influence of humans. Galápagos in many ways heads the list of places annually visited by tourists and conservationists who wish to encounter untouched nature and to follow in the footsteps of Darwin. Paradoxically, as modernity altered natural landscapes and created parks for projecting a pristine and people-free nature, these areas increasingly assumed natural conditions only through intensive human management. Almost 97% of Galápagos' terrestrial biome has been placed under the protectionist umbrella of an oft-criticized Yellowstone Model.[1] However, against the wishes of earlier conservationists who advocated for a completely people-free refuge, more than 3% was set aside for humans.

Galápagos has nonetheless been deliberately and passively shaped and influenced through various motivations and intentions since its early encounter with humans. The past 500 years of human history in Galápagos can be conveniently discussed in four periods (González et al. 2008).[2] It begins with a protracted period (1535–1832) of extractive exploitation, principally by seafaring explorers, buccaneers, whalers, and sealers. The islands served as a source for the removal of natural products from the land and surrounding ocean with minimal interest in environmental transformation or conscious recognition of any consequences resulting from extractive efforts. Seafarers used the islands as refuge and for replenishing supplies. Whalers came to exploit the bountiful waters for the oils and fats of marine mammals required by their industrializing worlds. Tortoises, sea lions (Galápagos fur seals), and iguanas were systematically harvested, some almost to extinction. Simultaneously, new organisms like rats, goats, and domesticated cultigens were both accidentally and intentionally introduced. Left alone to reproduce and multiply, they proliferated usually at the expense of endemic organisms.

Immediately after their accession by Ecuador, the islands were colonized by humans (1832–1959). In addition to establishing a human presence for the young republic, the main goal of colonization was to tame nature and convert island environments into humanized landscapes, which corresponded with expectations of production and development during an age of progress. Colonization of Galápagos by nineteenth-century entrepreneurs like Villamil and Valdizán on Floreana, Cobos on San Cristóbal, and Gil on Isabela, promoted the large-scale transformation of landscapes and the introduction of desirable plants and animals that reflected the practical and aesthetic perspectives of their mainland Ecuadorian homes and the broader world.

After the repeated failures of colonies on Floreana, the focus shifted to San Cristóbal and the burgeoning activities of El Progreso. As humans relocated, their abandoned but flourishing landscape capital reverted to enduring feral gardens that continued to serve island colonists with important "wild" resources. Landscapes are transformed according to the templates and aspirations that arrive with colonizing humans. Manuel J. Cobos acted on a shared dream of transforming nature through technology to create commodities and value. The late nineteenth century was characterized by a second industrial revolution that upheld industry, technology, and progress as a new secular religion allowing humans to transform and control nature. Electricity, fossil fuels, steel, and the telegraph were emerging as powerful new tools for transformation and development. Cobos was like his contemporaries—Edison, Tesla, Ford, Rockefeller, and mainland rubber barons—who were creating the foundations of a new capitalistic venture in a context where wealthy and powerful individuals could control human labor and landscapes to expand their vision and accumulate wealth by promoting progress and development through thought and action.

Landscape is transformed and shaped in different ways at different times. Whereas a later conservationist period considered Galápagos as harboring unique and pristine landscapes that required human protection, an earlier period of its history was guided by development, when nature was considered an alien reality that had to be controlled by humans. Cobos expanded his vision into an agroindustry. Reflecting the shared ideas of his time, the landscapes of San Cristóbal were shaped from top to bottom, and he and his associates introduced modern mechanical infrastructure to service extensive plantations and pastures. Historic sources and archaeological evidence attest to the significant modifications that transformed the endemic highland landscape into his vision of Hacienda El Progreso. Vegetation cleared for pasturage and plantations was used for construction and fuel. Exotic plants and animals were intentionally introduced and tended on a grand scale, whereas others were accidentally imported and subsequently proliferated. In the process, endemic organisms were locally extirpated[3] or left to compete with their new counterparts. Machines, engines, and other essential symbols of modernity were imported to this remote island from North America and Europe.

The humanization of landscapes persisted after Cobos's death in 1904. As the hacienda disintegrated into small farms, agriculture became the main economic activity in the islands between 1930 and 1970. None of this, however, equaled the fervor and level of organization achieved under Cobos's supervision. After the 1950s, many of his former workers relocated from the highlands to coastal settlements where fishing assumed increased importance and agriculture began to wane as the dominant economic activity. Colonization, agriculture, and fishing continued throughout the islands, usually with mixed results. Galápagos was largely left to the *colonos*, whose primary objective was to domesticate nature through domination and control during a period of isolated self-sufficiency, an ethos that to some degree extends into the present.

After Charles Darwin visited the islands in September 1835, and especially after the publication of his controversial *On the Origin of Species* in 1859, Galápagos was to become an important laboratory for the study of evolution and a location to collect specimens for study in centers of scientific knowledge and research. Expeditions to the islands, principally for the purpose of scientific collection, were a regular activity, dating back to the late eighteenth and extending well into the twentieth century (Epler 2013: ch. 13). Indeed, Darwin's celebrated visit was focused, not on conservation, but on collecting minerals, fossils, plants, and animals for scientific study in the museums of Europe.[4]

During the 1930s, Galápagos was attracting attention in global scientific discourse as naturalists were advocating for in situ protection of island biota, which was increasingly preferred over studying biological processes from museum collections (Hennessy 2017: 71). By 1936, Ecuador had designated most of the islands as nature reserves primarily to restrict hunting and extraction of resources; however, infrastructural support from the government was limited. It was only after the Second World War, through sustained lobbying by influential biologists,[5] that serious consideration was given to the development of a nature preserve, which became the Parque Nacional Galápagos (PNG) in 1959 to celebrate the centenary of Darwin's *On the Origin of Species*.

Over the ensuing 40 years (1959–1998), wilderness conservation was increasingly prioritized. Driven by environmental conservation and the interests of a nascent tourism industry, and capitalizing on the legend of Darwin, Galápagos was being converted into a natural laboratory within a pristine landscape (Quiroga 2009). From its earliest beginning, the park was inextricably twinned with tourism. Offered simultaneously with the dedication of a biological research station on Santa Cruz Island, the idea of ecotourism, or "controlled tourism that looks and collaborates, but does not destroy" (1961 editorial in *El Comercio*, quoted in Hennessy 2017: 86), would fulfill the dual goals of conservation and revenue generation. A 1966 recommendation for park administration advocated boat-based tourism, with customers sleeping on board and visiting restricted onshore areas accompanied by park guides. In 1968, the same year the PNG received its first personnel, the *Golden Cachelot*, a three-masted schooner with room for 10 passengers, was ready for biweekly adventure (Epler 2013).[6]

The original Emergency Law 17 of 1959, which set aside reserves of exclusive dominion by the state for preservation of flora and fauna, also exempted lands that at the time were in the possession of colonists (INGALA 2005: 51). In 1974, boundaries for the PNG and for areas of urban and rural settlement were designated, while land ownership for the latter was formalized (Villa and Segarra 2012).[7] In 1985, Galápagos was incorporated as a Biosphere Reserve under the UNESCO Man and the Biosphere program for resource conservation and the improvement of human/environmental relationships. In the following year, the newly recognized Galápagos Marine Reserve (RMG, Reserva Marina de Galápagos) immediately became one of the largest protected areas in the world. Over the ensuing years human population more than tripled in the islands as immigrants fled conditions on the mainland to pursue opportunities provided by the dramatic increase in tourism and fishing.

The current period, which attempts to strike a balance between development and tourism, was ushered in with the passing of the Special Law of Galápagos (LOREG, Ley Orgánica del Régimen Especial de la Provincia de Galápagos) in 1998, which became part of Ecuador's new constitution. The LOREG acknowledges historic interactions that existed between inhabited zones and protected areas, and seeks to promote the maintenance of ecological systems and biodiversity, especially of endemic biota, through sustainable and controlled development. It also emphasizes the privileged participation of local communities in developing a sustainable economy (INGALA 2005: 59). At the same time, the limits of the protected areas for marine ecosystems were extended, making it the second-largest marine reserve in the world.[8]

The *Colonos* Today

Galápagos environments experienced an early period of integration into the world economy. The paradigms and strategies for how they would be shaped were guided by the actions of international actors like buccaneers and whalers, and later through the aspirations of mainland colonists and entrepreneurs. The islands then shifted into a period of relative isolation largely under local control. A later period would once again integrate

Galápagos environments into the world economy and international influence through the growth of fisheries, science, and tourism.

Until fairly recently, the major focus of human population and activity throughout the Galápagos Islands was located in the humid interior highlands of larger islands where regular sources of water and sufficient soil development provided the necessities for longer-term survival. El Progreso had earlier established itself as the economic hub for the controlled flow of commodities between the islands and the rest of the world, particularly through its historic hacienda and the ambitions of its creator, Manuel J. Cobos. Following his death, the legacy of Hacienda El Progreso continued in various failed schemes, and eventually the town became increasingly dormant. Exportation of highland resources began to wane as most activities shifted to the lowlands. Increasingly, the economic boon afforded by fishing and tourism created a new demand for labor, which was realized in a demographic shift, especially of younger people, to the coastal towns. Local highland agriculture continued, but in the hands of an increasingly aging population.

Into the first half of the twentieth century, life on the islands was highly insular, with a largely subsistence economy based on hunting, collecting, and agriculture. It was characterized by local, community-based production and provisioning, informal barter, and cooperation and exchange between highland and coast, with nonlocal supplements provided through the infrequent visits of boats from the mainland. The family-based subsistence orientation began to change with declarations for turning Galápagos into a protected area and the accompanying opportunities afforded by the establishment of a park and its growing engagement in related commerce and tourism. Today, the majority of Galapagueños are employed in service, generally for tourism; only an estimated 9% of the population is involved in agriculture, ranching, forestry, or fishing (Salvador Ayala 2015). Agriculture is restricted today to the four inhabited islands, where it is practiced to varying degrees of intensity.

Although the coupling of ecotourism with nature conservancy might be offered as a sustainable way to balance the demands of humans while maintaining the integrity of endemic ecosystems,[9] it is not without its attendant problems. Tourism has generated direct and indirect impacts both on local resources and on the lives of Galapagueños. Land speculation is rampant; in some areas agricultural land is being converted into recreational space, communities are changed and displaced, and local food sovereignty and access to water are endangered (Salvador Ayala 2015: 60). Most of the income generated by Galápagos tour boats does not enter the local economy but remains in the hands of mainland or foreign operators; however, this is beginning to change with an increase in land-based tourism. Boat-based tourism can lead, directly and indirectly, to increased invasion by foreign exotics and increased pollution.[10] Ecotourism regularly contracts thousands of migrants to fill vacant positions, leading to "one of the few cases of domestic illegal migration in the world" (Brewington 2013: 115).

The effects of the new economy on local subsistence are both striking and predictable, as a food system installed by development and largely meant for tourist consumption has acted as a disincentive to local production (Salvador Ayala 2015: 85–86). Locally produced food is competitively disadvantaged in comparison to food imported from the mainland. Tourism demands a regular and dependable supply of high quality, inexpensive items, which are readily accessed through importation. In comparison, local products are often seasonal and erratic, of poorer quality, and more expensive.[11] This has a downward spiraling effect on local food production and increases the dependence of island populations on imported food.[12] Today, the roughly 19,000 ha of agricultural land in Galápagos is in the hands of only 755 farmers (Allauca et al. 2018: 14). Owing to its relative isolation, only Floreana is self-sufficient, whereas local agriculture contributes to as little as 9% of islandwide consumption (Salvador Ayala 2015: 69). Most of the demand is supplied by imports on boats and planes.

San Cristóbal, once the productive hub and exporter of fish and agricultural products on the islands, has been relegated to a minor role in the food chain and reliance on importations. According to the 2014 census, a total of 5,577 ha of the island ZUE is divided into 257 agricultural units (UPA, Unidades de Producción Agropecuarias).[13] More than 60% of farmers today have been on their lands for more than 10 years and range in age from 17 to 88, with a mean age in the fifties. Approximately 59% of the land is devoted to pasturage (Figure 30), commonly intercropped with citrus and coffee, for cattle that forage principally on exotic *Brachiaria*, *Pennisetum*, and *Panicum* grasses.[14] As many as 147 species are grown principally for food, especially oranges, plantains, limes, bananas, tangerines, avocados, yuca, papaya, and pineapple. Almost 90% of production is destined for local consumption, and not for sale (Allauca et

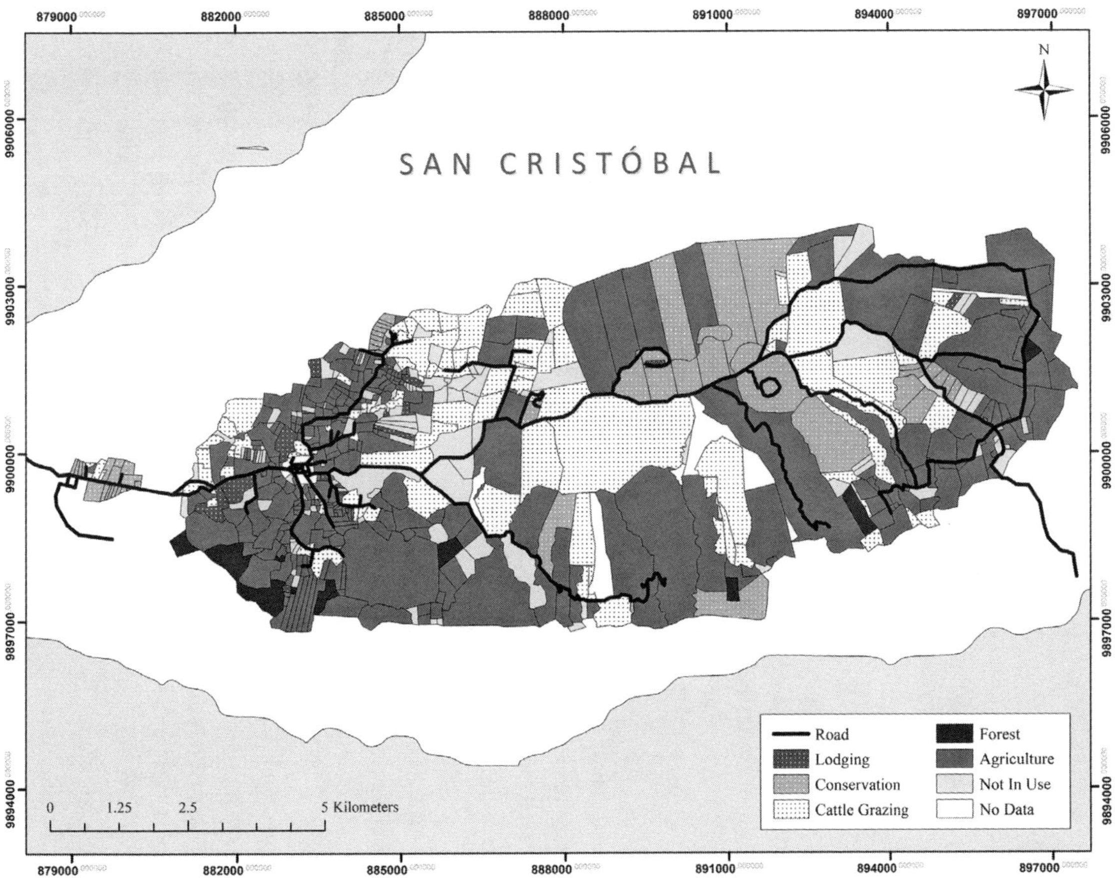

Figure 30. Land use in the Zone of Special Use (ZUE) of San Cristóbal Island (redrawn from MAGAP 2014). Map Datum: UTM 15S WGS 1972, created by Rylee Harlos.

al. 2018). Reduced productivity and land abandonment are locked into a downward spiral. Abandoned land encourages the increased dispersal of invasive species, which discourage future cultivation, leading to abandonment. A self-sufficient subsistence economy, a hallmark of previous populations, is incompatible with conservation, which has directly and indirectly forced its practitioners to abandon their livelihood and depend on imported commodities, while exacerbating exotic invasion.

Local food production in Galápagos is today afflicted by interrelated problems of human and natural origins. Recent interviews involving a representative array of stakeholders in the local food system on Santa Cruz and San Cristóbal islands reveal some shared concerns. Worrisome issues involving access to natural and human resources, the basic economics of local food production, a lack of governmental and local assistance, and problems associated with conservation goals, agricultural pests, and introduced invasive organisms are regularly articulated as major hindrances standing in the way of local food security (Brewington 2011; Kerr et al. 2004: 26; Laso 2016).[15]

Principal among these is water accessibility, a fundamental issue long plaguing the human experience in Galápagos, where water availability is today insufficient to meet human demand (CGREG 2016). As of 2010, an estimated 89.9% of San Cristóbal's population obtained water from the public network; however, an almost equal percentage of the population rated its overall quality as average to poor (CGREG 2016). The potable water plant in El Progreso, inaugurated in September 2013, receives its water from three principal sources, located in springs at Cerro Gato, El Platano, and La Toma (Becerra Chaves 2017). Other minor sources of water in the highlands include wells and springs on private lands.

Problems with water availability are less pronounced on San Cristóbal, where large and permanent sources in the interior highlands served as a principal reason for the existence of the historic Hacienda El Progreso. Nevertheless, farmers today generally report that water availability has decreased, and that the rainy season is increasingly less reliable than in past years. Conflicts occasionally arise among neighboring producers over access to water. Recent government efforts to improve agriculture in the highlands have included the construction of retention ponds and the introduction of techniques like drip irrigation (Laso 2016).

Despite the relatively enhanced water situation on the island, the possibilities of future problems loom large. In the shorter term, the demands of increased land-based tourism, of which most Galapagueños are understandably in favor (Brewington 2013: 115), will certainly tax the current water supply, especially on the coast. Over the longer term, the possible effects of global warming on local precipitation patterns are worrisome. Although it remains uncertain whether future weather patterns might increasingly resemble El Niño or La Niña conditions, the very real possibility for an augmented frequency of extreme weather events, and extended droughts, is vexing (Cai et al. 2014; Vecchi et al. 2008).[16]

El Progreso continues to be a popular venue for Sunday outings by island residents, when the town comes alive with soccer matches, picnics, and visits to the restaurants on its outskirts which open just for weekend visitors. Many Progreseños own residences or property on the coast yet remain strongly attached to their highland home. Nevertheless, as younger people increasingly eschew farming in favor of lucrative nonagricultural jobs in fishing, tourism, and public services, the demographics of resident food producers is progressively aging.[17] Shortage of labor is a very real problem. Although the average gross monthly wage is approaching US$450 nationally, the LOREG not only restricts migration of potential laborers to the islands, but it also guarantees that wages for Galapagueños must be at least 75% higher than those on the mainland as a concession to the higher cost of living (Kerr et al. 2004).

The economic viability of farming is further hindered by interrelated factors that include the poorer quality of, and general lack of demand for, local products, coupled with competition from similar products imported from the mainland. Local products are sold mainly in the central market of Puerto Baquerizo Moreno (CGREG 2016: 103), but a common remark by restaurant owners and experts in Laso's (2016: 5) sample mentions their lack of quality compared with imported products. The inability of local producers to regularly supply quality products at high volumes for the constant demands of the tourism industry increases a reliance on imported products. Since 2000, maritime cargo shipments have quadrupled, and in 2011, 803.4 tons of fresh food were being imported from the continent on a monthly basis (Bigue et al. 2012: 15).[18] Even with the cost of transport, imported products are usually available at equivalent or cheaper prices than their local counterparts.

Producers, in particular, have cited a number of problems associated with a general support for agricultural pursuits, including lack of funding, infrastructure, training, collaboration among farmers, and adequate transport for getting products to market. They also complain about labor regulations that make it difficult to hire local workers and to import laborers from the mainland. High interest rates and long processing times for loans hinder access to credit; also funding, particularly for aging farmers, is reported to have worsened over time (Laso 2016). The federal Ministry of Agriculture (MAGAP, Ministerio de Agricultura Ganadería Acuacultura y Pesca) is attempting to assist highland farmers through the creation of larger water reservoirs, the introduction of different techniques like drip irrigation, and experimentation with new crops.[19] MAGAP and the National Institute of Agricultural Research (INIAP, Instituto Nacional de Investigaciones Agropecuaras) are instituting a plan for bioagriculture, defined as "agriculture adapted to island conditions that generates a supply of healthy food and strengthens government actions to conserve protected areas" (CGREG 2016: 103; also Allauca et al 2018).

Although a large majority of Galapagueños understandably wish to preserve their island lifestyle, an even larger proportion of the population agrees with the need for island conservation (Brewington 2013: 111). Nevertheless, enduring transformations of the landscape to accommodate agricultural production have altered endemic plant and animal habitats and introduced exotic organisms, many of which have become invasive. Farmers are well aware that the numbers and geographic spread of invasive species have increased in recent years. Their fears are strongly supported by data that show a dramatic increase of invasive species over a period of just 19 years (Villa and Segarra 2012).[20] They also report of enhanced problems created by agricultural pests that contribute to a reduction in both product yield and quality.

The spread of invasive species, the increasing difficulty of farming in the highlands, and the abandonment of agricultural plots are locked into a vicious cycle as abandoned land enhances the propagation of invasive species, increasing the difficulty of farming. Recent figures show that almost a quarter of UPAs throughout Galápagos' ZUEs are abandoned and generally covered in invasive species and/or trees, or are destined for housing (CGREG 2016: 98). On San Cristóbal Island, aggressive blackberries (*Rubus nivea*), guava (*Psidium guajava*), and rose apple (*Eugenia malaccensis)* are the principal invasive species which have spread rapidly over large areas (Villa and Segarra 2012). About 70% of the plant species in the highlands are now introduced, and many are invasive, including Spanish cedar (*Cedrela odorata*), elephant grass (*Pennisetum purpureum*), purple coraltree (*Erythrina fusca*), and quinine (*Cinchona pubescens*) (Laso 2016). Today exotic species cover thousands of hectares in the highlands where they affect the abundance, distribution, and survival of their native counterparts.

A high degree of unpredictability can accompany coupled human-natural systems as the interaction between introduced and native species is often complex, unpredictable, and dynamic. Some invasive species create conditions that contribute to an increase of endemic organisms through the creation of new niche opportunities; others compete either indirectly through habitat invasion and nest destruction or directly through predation. While most farmers do not report having any conflict with wildlife, they can have mixed reactions to both endemic species and exotic invasive species, primarily because of their focus on agricultural production.

Farmers may not consider invasion by exotic species an ecological problem, but they are well aware of the agricultural problems they can present. Encroaching invasive species like blackberry or guava are quick to colonize agricultural land, thereby elevating farming effort; others like rose apple and quinine might require farmers to use restricted herbicides, which increases costs. Invasive rats and insects are major problems for highland farmers, who are in agreement with the park that they should be eliminated.[21] However other invasive species like feral goats and pigs are hunted for food, and their elimination for the sake of saving endemic species may not be considered justifiable. Of course, invasive domesticates like cats, dogs, goats, cattle, and pigs remain desired farm fixtures. Guava provides food and shade for humans, economically useful coral tree and cedar are preferred fencing and building materials, and highly invasive elephant grass is important as forage for cattle ranchers (Laso 2016).

The relationship between humans, other invasive species, and endemic species is also context dependent. Many producers, especially ranchers, are neutral or welcoming to tortoises, even though they can compete with cattle for forage. However, most ranchers prefer elephant grass, which can impede tortoises on their annual highland migrations, a pattern that in turn might be changing because of their attraction to guava fruit (Blake et al. 2015). Fences erected by farmers to keep out tortoises reportedly attracted by the smell of crops may also impede migration. One project is under way by the PNG to promote tourism by introducing tortoises from breeding centers into farms where participating farmers have built appropriate fencing.

Charismatic species like the endemic hawk were extirpated on San Cristóbal to protect chickens, and native finches (*Geospiza* spp.) are viewed as pests because they eat seedlings and fruits (Laso 2016). Habitat fragmentation through human activity regularly alters the frequency and distribution of endemic species and in certain places has led to local extinction.

The *Colonos* Tomorrow

San Cristóbal had been the most populated and intensively exploited of the Galápagos Islands since the early period of colonization and continuing up to the years of the tourism boom. In many ways, it is today the most denigrated of the populated islands. Floreana harbors a vestigial enclave of hardy residents, and Isabela is romanticized as a distant frontier. Santa Cruz, populated by foreigners, privileged Ecuadorians from mainland urban centers, and conservationists, is upheld as the progressive center for conservation efforts and tourism. In contrast, it is not uncommon to hear San Cristóbal referred to as a hopeless mess, a lost cause, or a ruined island. It is often disparaged as a working island populated mainly by poorly informed people who trace their origins to rural areas on the mainland, and whose major activities often conflict with progressive international perspectives.[22]

These characterized images of San Cristóbal are directly linked to the island's historic role in human colonization and specifically to the Hacienda El Progreso. The abandonment of San Cristóbal's highland agricultural legacy is the result of numerous factors, which are underlain by globalization. Many residents are concerned about the future of highland agriculture, and even though most livelihoods have shifted to the coast, a lingering connection to the interior *montañas* remains strong. In many ways the memories and experiences of highland life have colored the narratives of being a true Galapagueño. They include a time of individual hardship and plenty with no diseases or pests, when nature ruled daily life and a simpler existence created a sense of belonging. The smells, sights, and sounds of the highlands are a strong social referent for a people who see themselves as locals, and whose long history bestows a special right to the islands.

The realities of this idealized highland lifestyle changed dramatically as focus shifted to conservation, wildlife protection, and the management of nature. Connected responsibility was transferred to the external authority of governmental and nongovernmental organizations whose protective measures assumed control of transforming the local environment and generating a pristine Galápagos. Today, initiatives are under way to integrate San Cristóbal's agricultural zone into the plans, programs, and projects of conservation. Novel ecosystems are being embraced by some for their high degree of agricultural diversity, which is important in the Bioagricultural Plan for Galápagos (Allauca 2018: 39).[23] In order to encourage agricultural production adapted to island conditions, governmental plans seek to promote agroecology and sensitize consumers to responsible consumption, fortify agricultural infrastructure, implement irrigation along with technical assistance and credit, and modernize and improve processing chains (CGREG 2016: 105).[24]

The long history of fishing, an activity in direct opposition to conservation efforts of the RMG, continues today with the pursuit of sea cucumbers, lobsters, and fish for local and export consumption.[25] In accordance with the LOREG, which promotes the comanagement of fishing zones, quotas, permits, licenses, and chain of custody by various stakeholders, the government is attempting to align fishing with the goals of conservation and local sustainability. Increased support for additional remote technology, boats, and planes is being introduced in efforts to ramp up protection of the RMG in order to curtail illegal activities, especially by large industrial fishing vessels from mainland Ecuador and foreign countries. Government is also empowering local fishermen with greater control over pricing, through instituting best practices certifications and *Sello Galápagos* branding, which assures consumers that their purchase is of high quality, legal, and a sustainable catch from socially responsible cooperatives (CGREG 2016).[26]

As Galápagos has transformed into a touristic center that increasingly generates wealth for towns on the coast, highland tourism is seen by many as a possible entry into the new economy and a way to generate income for local inhabitants. El Progreso has made various efforts to increase its relevance for tourism. Many infrastructural improvements have been recently added, with the completion of the paved road, a new park, upgraded amenities, and public buildings. Some more traditional forms of tourism have met with variable success, and newer venues for voluntourism and agrotourism have been established by local, mainland, and foreign entrepreneurs.[27] The ruins of the historic hacienda linger as a possible attraction. In years past, the Spanish Agency of International Cooperation (AECI, Agencia Española de Cooperación Internacional) initiated a number of tourist-oriented improvements to the house site, and occasionally tour buses would stop; however, this rarely happens today.[28] Those tourists who venture into the highlands to visit the crater lake at El Junco, the tortoise hatching station at La Galapaguera, or the beach at Puerto Chino usually bypass El Progreso; however, the majority of tourists who make the pilgrimage to Galápagos tend to remain on the coast or tethered to their tour boats.

Although there have been times of greater and lesser association with the world beyond its isolated setting in the eastern Pacific, Galápagos has always been a globalized place. Today, it is fully integrated into an international economy, which continues to create challenges for local conditions. Going forward, the future of the archipelago lies in the hands of various stakeholders, many external to the islands. On a local level, González et al. (2008) identify the adoption of an *isleño* lifestyle as a key to building socioecological resilience and initiating the archipelago's transition toward increased sustainability. It involves disconnecting some of the strong linkages to the mainland which dominate an *insular* lifestyle by developing an understanding of Galápagos as a special place. It requires consumption patterns that differ from those of the continent, prioritizing local resources, and adopting management and extractive practices based on island realities. For this to work, it must incorporate an interisland collaboration by an *isleño* population which emphasizes local identity and a marked sense of place based upon a lengthy shared history.

Galápagos and Galapagueño culture are in a process of transition from a production-based economy to one based on services, especially tourism. This transition has resulted in a less polarized relationship between conservationists and local people as a new hybrid paradigm is emerging (Quiroga 2013). The future of San Cristóbal's agricultural sector is unclear, particularly as more people are abandoning their fields, more invasive species are taking over, and farmers are aging. This transition presages that unless new financially and culturally successful options emerge, the future for sustainable agriculture will be difficult.

NOTES

Chapter 1. Introduction

1. Urban Zones include the four major towns of Puerto Baquerizo Moreno (San Cristóbal), Puerto Ayora (Santa Cruz), Puerto Velasco Ibarra (Floreana), and Puerto Villamil (Isabela). These towns also include: sites of distribution and assembly (e.g., ports), sites of special service (e.g., airports), and sites of administrative and logistical infrastructure (INGALA 2005).

2. Terminology follows Pyšek et al. (2004) for plants: *native*—indigenous to an area without human involvement; *alien*—exotic due to intentional or unintentional human involvement; *casual alien*—occasionally reproducing outside cultivation but dying out unless repeatedly introduced (e.g., Duffy 1981; Phillips et al. 2012 for examples of casual alien animals); *naturalized*—an alien establishing self-replacing populations for a minimum of 10 years; *invasive*—a naturalized alien self-reproducing in large numbers over larger areas; *transformer*—an invasive alien changing "the character, condition, or form or nature of ecosystems over a substantial area" (Pyšek et al. 2004: 136).

3. Placed on the UNESCO list of World Heritage Sites in Danger in 2007 in an attempt to mobilize support for their conservation, the Galápagos Islands were removed from the list in 2010, owing to the Ecuadorian government's subsequent progress in combating the threat of invasives, rampant tourism, and overfishing. http://whc.unesco.org (accessed March 20, 2018).

4. Darwin visited Galápagos between September 15 and October 20, 1835, toward the end of the HMS *Beagle*'s five-year voyage. During this time, he spent 19 days on San Cristóbal (Chatham), Floreana (Charles), Isabela (Albermarle), and James (Santiago) islands. Although his acceptance of the mutability of species and concept of natural selection is often attributed as a conversion that suddenly materialized during his brief sojourn on the islands, these ideas emerged after his return to England in 1836 and in association with subsequent study of his collections by eminent colleagues. The coupling of a "eureka moment" for Darwin with his visit to Galápagos surfaced only later in the twentieth century with efforts to conserve the islands as an "outdoor biological laboratory" in the 1930s, and with the later beginnings of "controlled tourism" in the 1960s (e.g., Durham 2012; Hennessy 2017; Hennessy and McCleary 2011; Quiroga 2013; Quiroga and Sevilla 2017; Sevilla 2017a; Sulloway 1982a, 1982b).

5. Over the last decade, the number of alien organisms introduced into the islands is estimated to be as high as 1,500. They include between 600 and 870 plants, many of them ornamentals, with 30–45% having become naturalized. As many as 536 alien invertebrates include 463–500 insects (almost doubling a 1998 inventory) comprising almost one-quarter of the insular insect faunas. Approximately half of the 44 introduced alien vertebrates, including 17 mammals, 12 birds, 9 reptiles, 4 amphibians, and 2 freshwater fish, have become naturalized. The estimate also includes at least 10 diseases (Atkinson et al. 2012; Causton et al. 2006; Gardener et al. 2010; Guézou et al. 2010; Phillips et al. 2012; Soria et al. 2002). Active conservation options include containment, long-term control, and eradication (Atkinson et al. 2012). Major eradication projects of larger mammals have included the removal of 41,682 goats from Pinta Island (Campbell et al. 2004); 1,523 donkeys and 89,474 goats from Santiago Island, and 1,212 donkeys and 62,818 goats from northern Isabela Island (Carrion et al. 2007, 2011; Lavoie et al. 2007); more than 18,800 pigs from Santiago Island (Cruz et al. 2005); and rats from Rábida, Pinzón, and smaller outcrops and islets (Nicholls 2013).

6. The parameters of the PNG management plan (INGALA 2005: 42–46) explicitly recognize Galápagos as both a terrestrial and marine ecoregion, and a socioecosystem formed through a process of colonization and transformation early in Ecuador's Republican period and very rapidly since the 1980s. Both systems are interactively connected; therefore, economic, social, and cultural circumstances of Galápagos must be included in park management. Conservation and development in Galápagos are integrated from the perspective of the socioecosystem as scientists and local Galapagueños act cooperatively to model this integrated response to development of the protected ecosystems.

Chapter 2. Humans Encounter Galápagos

1. A comprehensive alphabetical listing of island names can be found in Woram (1989), and at http://www.galapagos.to/Table.php. Unless specifically stated, we follow commonly used names for islands today, which tend to follow the Ecuadorian naming protocol of 1892 with a few exceptions.

2. *World's End* (Beebe 1924); *Clinker Islands* (Otterman 1983, 1993); American military personnel nicknamed Baltra Island *The Rock* while stationed at Base Beta during the Second World War. After annexation by Ecuador in 1832, some islands were used as penal colonies for political prisoners and the most hardened criminals; at different times Galápagos were contemplated for use as collateral, repayment of war debt, outright sale, rent, lease, annexation, and *res nullius* (Latorre 2011; Woram 2005).

3. In addition to numerous personal accounts of life on the islands, a number of general treatments, some presented from specific perspectives, are available. Perhaps the most authoritative histories have been written by Octavio Latorre. These are available through various venues in Ecuador, and recent privately published books include *Historia Humana de Galápagos: Nuevas Investigaciones* (Latorre 2011) and the seventh and English version of *The Curse of the Giant Tortoise* (Latorre 2012). Other Ecuadorian sources include Bognoly and Espinosa (1905), Borja (1948), Vanegas León (1998), and Idrovo (2005; 2008). John Woram has published for years on human history, and most recently *Charles Darwin Slept Here: Tales of Human History at World's End* (Woram

2005). He also maintains a comprehensive website (Woram 2016). Other English-language publications covering various facets of human history in Galápagos are available in paper and digital formats (e.g., Bassett 2009; D'Orso 2002; Egnal 2015; Epler 2013; Grant and Estes 2009; Lundh 1999; Otterman 1983, 1993; Slevin 1959; Von Hagen 1949).

4. Peter Martyr D'Angheira was appointed chronicler of the Council of the Indies in 1520 and collected documents and accounts on New World discoveries that were arranged in 10 chapters or "Decades" for Charles V before his death in 1526. Decade 3 was first presented to Pontiff Leo X in 1516 (MacNutt 1912[1530]).

5. See Estrada (1988), and later 1813 mention by Porter (1822: 123), with descriptions in 1684 (Dampier 1697) and 1823 (Morrell 1832: 120) from the Peruvian coast.

6. They include the 1953 Norwegian Archaeological Expedition, the 1954 Walt Disney Galápagos Expedition, the 1963 Geological and Archaeological Field School from the Escuela Politécnica del Litoral (Guayaquil) by Raul Maruri, and the 2005 Australian National University Expedition (Flett and Haberle 2008).

7. The "Ys de galapagos" appear for the first time on a 1530 vellum; however, Woram (2005: 10–12) demonstrates how these were later additions dating to 1561. At this early time the islands were also referred to as "Las Huérfanas" and "Las Encantadas," perhaps because they were difficult to locate or would appear or disappear on the horizon (Bognoly and Espinosa 1905; Donoso 2012: 220; Jiménez 1891; Vanegas León 1998). This explanation was also provided by Capt. Peralta, taken prisoner in Peru by William Ambrosia Cowley in 1684, after which he used the name "Isla Encantada de Cowley" (Donoso 2012: 220). Porter (1822: 177) attributes the Spanish Las Encantadas to the great difficulty that ships found in escaping them.

8. Drake was able to attack otherwise unsuspecting and unprotected prizes along the coast in part because of the common misconception that the Atlantic-Pacific passage had been blocked by a natural cataclysm (Clissold 1954: 107). With the backing of papal edict, the Spanish Crown invoked the old theory of *mare clausum* (closed sea) for their remote "Spanish Lake" in the Pacific during this early period of the colonial experience (Schurz 1922).

9. These earlier incursions into the Pacific were mainly affairs of the state; however, after 1643, intrusions were undertaken by pirates, freebooters, or buccaneers (Gerhard 1960: 137). "Buccaneers" may derive from "boucaning" or drying cattle hides, and boucaniers engaged in their illicit traffic (Eyer 1918: 6). It may also derive from Carib *bucan* or grill for smoked meat, and for persecuted populations who raised cimarron cattle in the Caribbean, selling meat to passing ships, and allying with them to capture the western Caribbean. Freebooter derives from the Dutch *vrie buiter*, from English free booter, or from fly boat, for the fast boats they used (Donoso 2012: 14).

10. "June the sixth, we set sail from Quiblo in the afternoon, bound for the Gallipagoes, which are over seven islands that lie under the aequinoctial, and about 100 leagues from the main" (Ayres 1684: 18). Donoso (2012: 17–19) suggests that Sharp knew of the islands from old maps captured from galleons and references to them from Spanish prisoners, before his return to England in 1682.

11. In order of appearance: King Charles's (San Cristóbal), Crossman's, Brattle's (Tortuga), Dean's (Pinzón), Eure's (Genovesa), Dassigny's, Bindlo's (Marchena), York's

(Santiago), Norfolk's (Santa Cruz), Narborough's (Fernandina), and Cowley's Enchanted (Cowley 1699; Woram 1989).

12. Donoso (2012: 241) claims that when Davis returned in 1687, the marmalade jars had been destroyed by the Spanish.

13. A version of the original document is found in Ducéré (1895), republished in Bradley (2015). Unfortunately, Ducéré's transcriptions for the entries in question appear to be summaries with inaccurate dates. Our transcription of the handwritten journal entries for July 18 and 19 in Massertie (1601–1700: 121) are as follows:

"Ce jourday[journee?] Samedy 18 du J. nous avons Envoye a terre pour Cherger [de?] Leau, [Lon?] Non ast point trouve aux Sourses qui sont en le port Le canost est [a le?] a unnes sourse qui est sous le vant [La I.?] port ils a L. Est[n] [?] yls ont trouve un peu Deau qui [Lith ille Du?] morne apres midi il y ast retourne pour voir Sy Lon[¿] pouvait y en faire Sa[¿] a minuit Le Canot est retourne a bord ils nont fait que troix chevres L'eau En touste La Journee. Ce Jourday Diman[e] 19 Du J. nous avons ramasse Les chevres qui Estoit[?] a terre Du I. Carenache."

"This day Saturday 18 of J[uly]. we went on land to look for water that we [did not find?] a source which was [supposed to be?] in this port. The canoe/boat [found? Went to?] a source which is below/under the port where it stopped and they [?] found a bit of water [that this island . . ? . .] mountain afternoon and came back to see [how far?] it could be to do it [?]. At midnight the canoe returned on board they had nothing except three goats the water all day. This day Sunday 19th of J[uly]. we gathered the goats that Were on land on I. Carenache [Careening Island]." Bradley (2015) and Donoso (2012) list various French names bestowed on different islands around this time: Mascarin (Española), Notre Dame, Île de Carenache de Flesingais (Floreana), L'île à Tabac, Île Brûlée (Isabela), L'île Vert (Santa Cruz), and Sain-Bernard, Saint Bernabé (Santiago).

14. By the mid-eighteenth century, increasing intolerance of piracy, coupled with the prosperity afforded by international trade and increased smuggling alongside the growth of naval protection, effectively curbed its practice (Gerhard 1960; Latimer 2009; Marx 1992).

15. As many as 100 barrels of high quality oil, one-third derived from the cranial spermaceti organ (junk or melon), was harvested from a large sperm whale. It was used for lubrication, candles, illumination, leather tanning, woolen cloth, soaps, paints, varnishes, shipwork, and cordage. Whale processing refuse was used for glue and fertilizer, and bone for stays, parasols, umbrellas, upholstery, trunk and bag framework, fishing rods, driving whips, and carriage parts (Tower 1907: 94–96).

16. Colnett (1798) named some islands, including Chatham (San Cristóbal), Hood's (Española), Barrington (Santa Fé), Duncan (Pinzón), and Jarvis (Rábida), and produced a chart that accompanied ships into the subsequent century.

17. This can be attributed to upwelling of the fast, eastward-flowing Cromwell Current or Equatorial Undercurrent that fertilizes surficial waters, resulting in the highest concentrations of phytoplankton west of Isabela and particularly Elizabeth Bay (Jimenez 1981). Galápagos were much preferred by whalers as they were relatively easy to get to, remote from continental turmoil, close to the "offshore grounds" of sperm whaling in the

Southeast Pacific, and the only landfall within the area from which needed provisions could be acquired (Epler 2013).

18. Tortoises, which Porter (1822: 162) referred to as "Gallipagos mutton," were taken from at least five of the largest islands in addition to Española, Pinta, Pinzón, and Santa Fé. Ships often sent boats to capture smaller, 50 to 75 lb. tortoises that could be transported by one man "backing down" with his prize. In coastal areas, smaller females were particularly exposed, which led to rapid population decline (Epler 2013). Larger tortoises were butchered, dragged by ropes, or slung to oars (Townsend 1925: 59–63). Townsend's (1925: 57) study of logbooks from 79 whaling ships suggests that 13,013 were captured during 185 visits between 1831 and 1868. He considers this figure "a mere fraction" of the total catch. Latorre (2011: 45) calculates that approximately 1 million were taken between 1788 and 1864 alone. Tortoises and marine turtles were also directly captured to supply the California market as early as 1850, particularly during the Gold Rush (Conrad and Pastron 2014).

19. Bahía de los Correos, or Post Office Bay, is today a well-known tourist site on the northern coast of the island. It began as a barrel erected by whalers to transmit letters while at sea.

20. Various islands were refuge spots for mariners desiring to desert ship (Epler 2013: ch. 7). Successful havens would certainly require some persistent water source. Encounters with marooned sailors were commonplace from Española in the east (Grant and Estes 2009: 99) to Fernandina in the west (Porter 1822: 142).

21. A crew member bestowed the honorific name of "Porter's Island" on the previously unknown Santa Cruz (Porter 1822: 165).

22. The goats consisted of a young male and three females, all of "the Spanish breed" except for one of the "Welch breed" and "a Peruvian ram with five horns" (Porter 1822: 237).

23. Latorre (2011: 56, fn. 14) suggests that it remains in doubt as to whether goats were introduced into these islands, as Villamil's 1861 testament uses the word "lanar," which applies to wool and sheep; however, Coulter (1845: 103) clearly describes "several groups of reddish coloured goats" during his 1833 hike in the highlands of San Cristóbal (see endnote 26).

24. Referring to the Pacific Ocean during his 1833 sojourn, Coulter (1845: 65) writes: "there is scarcely an uninhabited island in those seas, in the thoroughfare of shipping, on which there is a fertile spot of earth with a supply of fresh water, that has not its Robinson Crusoe on it." Two years later, Darwin (1845: 376) encountered men who had been sent from Floreana to Santiago in order to dry fish and salt tortoise meat, as well as a "hovel" at 2,000 feet in elevation for those employed in catching tortoises. These were in all likelihood the workforce sent out by Lawson.

25. Wheat, barley, orange, pomegranate, fig, papaya, banana, lemon, coconut, corn, sugarcane, and exotic cotton hybridized with endemic cotton (Grant and Estes 2009: 139) can be added to previous mention of pumpkin, potato, melon, and plantain.

26. "I was very agreeably surprised to discover on the island, grazing about near the stream, several groups of reddish coloured goats; they appeared from some cause or

other very wild, and dashed off into the bush on seeing me; however, the gun was too quick for them, and I shot one whenever I liked. They must have been left on the island a long time ago, and increased rapidly and unknown to all, for there is not any report of their existence on this island ever made; indeed they were no fools; they chose the best, most fertile, and unfrequented part for their homes. I have not met with any of them along the coast." (Coulter 1845: 103)

27. He claims that the men were removed from the island by Capt. Steivers of the London whaling ship *Favourite*. Here, Coulter is most likely referring to Capt. William Stavers ("Steivers" or "Stivers" could be a Cockney pronunciation of Stavers, who was in the west Pacific and Galápagos as early as 1808; Richards 1993, Clayton 2014: 188). Capt. Stavers was issued Letter of Marque with 14 guns (Clayton 2014) on the London whaler *Seringapatam*, captured by Capt. David Porter after being spotted off Bank's Bay on July 12, 1813. Stavers and another captured captain were given the whale-ship *Charlton* to transport, under oath, Porter's prisoners to Rio de Janeiro on July 19, 1813 (Porter 1822: 203–205). There was a Nantucket-based whale-ship *Favorite* operating in the Pacific in the early 1800s (American Offshore Whaling Voyages, http://integratedstatistics.com/ejosephs, accessed November 29, 2016), and a later Nantucket Capt. John R. Stivers, from 1859 to 1876 (Federal Writers Project 1987). The remains described by Coulter most likely predate the colonization of Floreana by Villamil, and so must the introduction of goats to San Cristóbal. It is possible that someone like Nicholas Lawson introduced them in the years prior to Asilo de la Paz, but if Coulter's claims about the three men, and Stavers's identification are accurate, then their introduction may have taken place before July 1813.

28. Briones and his accomplices sailed on to the Gulf of Guayaquil, where they killed 20 men aboard a naval sloop before their eventual imprisonment and execution (Latorre 2012: 42–43). The story served as the basis for a novel by Chilean writer Manuel Bilbao (1865) who, for a number of years, was a political exile in Guayaquil. It was subsequently republished in Ecuador (Webster 1904: 12–96).

29. One venture, the "Empresa Agrícola Peruana de Chatam," was reported to have been formed with a North American Capt. Norton (Webster 1904: 8).

30. Orchilla (orchil, archil) was used in the production of a plant-based dye (orcein, purple cudbear, lacmus). In Europe, the most esteemed form was made from lichens growing in the Canary and Cape Verde islands. The milled plant produced a fine powder that was treated with stale urine, soda, and lime water or gypsum to form a soluble "crimson bordering on violet" infusion used in wine thermometers and for dyeing marble, wool, and silk (Berthollet and Berthollet 1841). Currently, five species of the cosmopolitan genus *Roccella* are considered valid for Galápagos: four are endemic "false orchil" and *R. gracilis* (syn. *babingtonii*) is the indigenous but not endemic "orchilla," "orchil," or "Canary weed" found today on most of the larger islands except Fernandina and Marchena (Galápagos Species Checklist http://darwinfoundation.org, accessed Dec. 1, 2016). Wolf (1887: 12) describes it growing exclusively in the dry lowlands below 100 masl on rocks and trees directly exposed to onshore winds.

31. He lists sugarcane, cassava, sweet potato, potatoes, cotton, indigo, lettuce, cabbage, radishes, carrots, beets, artichokes, plantains, bananas, avocado, oranges, lemons, figs, and recently introduced palms (Wolf 1887: 15).

32. Thomas Levick (possibly "Lewis" in Ecuadorian pronunciation), aka "Johnson," claims to have let goats loose on Española in 1897 (Woram 2005: 221). Mann (1909: 26) visited the island in 1907, describing it as a "miserable place" with goats subsisting on cactus and water as its sole inhabitants. Levick figured prominently in the tragedy of Valdizan's assassination by organizing his men to resist the criminals and dispatching two of them. He remained on Floreana to continue the colony's business interests, before eventually relocating to San Cristóbal (Latorre 2012). Population estimates for resident horses, donkeys, pigs, cattle, and goats on Floreana, San Cristóbal, Santiago, and Isabela were also provided during the brief British zoological collection trip on the HMS *Peterel* in 1875 (Salvin 1876: 456).

33. Chemists have been exploring the chemical structure of orcein for 150 years, and by the second half of the nineteenth century Austrian scientists had synthesized Resorcin (Hlasiwetz and Barth 1864).

34. On his return voyage from Isabela, Mann visited Santiago and hiked into the highlands, where he reports a "large pig and several fine sleek donkeys besides numerous signs of others" but no cattle, goats, or dogs (Mann 1909: 68).

35. Listed crops included: sugarcane, cassava, otoy (*Xanthosoma*, cocoyam), yams, maize, potatoes, cabbage, beans, melons, plantains, eucalyptus, ceiba, oranges, lemon, papaya, and coffee. Gil had converted coastal lagoons into palm orchards where figs, plantains, papaya, and breadfruit were cultivated (Mann 1909: 46; Martínez 1915: 79, 86–89).

36. Mann (1909: 52–53) notes that some of the cattle had docked tails resulting from encounters with dogs, and that despite having eradicated hundred of dogs with poisons, there was no visible decrease in their population. Plenty of tortoises were seen by both authors, but Mann predicted their extermination, along with cattle, by dogs. José Villamil may have first introduced dogs onto Isabela in 1835 when he abandoned hounds on the southern slopes of Sierra Negra that had accompanied him on a hunting trip. Genetic and morphometric analyses of extirpated feral dogs have established that these populations were derived from a single founding population that had survived through predation on marine iguanas, juvenile sea lions, feral cattle, and scavenged prey trapped by hunters on the island for almost one and a half centuries (Reponen et al. 2014).

37. The French had abandoned their efforts to build the canal in the previous century, and it was in part because of their earlier activity and secretive reports about American interests that the British Admiralty had sent the HMS *Herald* and *Pandora* into the Pacific for their eventual rendezvous with Galápagos in 1846 (Samson 1999).

38. This period is recorded in numerous firsthand narratives that describe day-to-day life, including details of introductions, the status of invasives, and ecological transformation on San Cristóbal, Floreana, Santa Cruz, and Santiago (e.g., Angermeyer 1989; Conway and Conway 1947; Lundh 2006a; Ritter 2015[1931]; Strauch 1936; Wittmer 1961).

39. Joseph Slevin (1959: 122–125), a member of the 1905–1906 California Academy of Sciences expedition aboard the schooner *Academy* lists the various scientific expeditions to Galápagos between the World Wars. Woram (2016) provides an annotated list and accompanying images of ships that had visited the islands over the years, http://www.galapagos.to/SHIPS/INDEX.php

40. An early attempt by La Compañía Colonizadora Suizo-Escandinavia to attract Swiss and Scandinavian settlers went nowhere; however, Norwegian interest was piqued in 1907 by the tragedy of the *Alexander*, which was abandoned while transporting coal from New South Wales to Panamá. The captain and some crew arrived safely on San Cristóbal but ten sailors aboard the vessel were presumed dead. The wreck was subsequently located on Isabela at Iguana Cove, and later, eight survivors were located on Santa Cruz (Lundh 1999; Hoff 1985).

41. The station at Post Office Bay was to be used as a base for exporting dried fish and lobster, sperm whaling, fur seal hunting, exploiting wild cattle, growing tropical fruits, salt, lime, and sulfur extraction, seaside tourism, sport fishing, and provisioning ships passing through the Panama Canal (Lundh 1999; Hoff 1985).

42. Alf Kastdalen (1982), a longtime resident of Santa Cruz, maintained that in 1923 Capt. Rafael Castro released a bull and two cows in Academy Bay before sailing off. He also mentions that pigs, descended from wild black Floreana pigs and an improved white mainland breed, were released in 1927 after a Norwegian stock-raising attempt was abandoned. He claims that milking goats were introduced to the island after 1925 from Santa Fé, Baltra, and Santiago.

43. The wreck of an Austrian ship *Turul*, renamed *Carawa*, was leaking its load of kerosene on a shoal in the middle of the entrance to Wreck Bay when they arrived (Hoff 1985). In 2001, the Ecuadorian supply ship *Jessica*, loaded with diesel and bunker fuel, hit the same reef. http://www.galapagos.to/SHIPS/INDEX.php

44. Two families, in particular, were successful as they relied on tropical crops. Karin Guldberg, raised on the island by her widowed father Thorleiv, eventually married and divorced Cobos, raising three daughters and three sons. She began to raise cattle and by 1965 had nearly 600 head. Later, with her sister Snefried, they possessed 1,200 to 1,300 head on their ranch Pampa Mía (Lundh 2006a).

45. After the Second World War, Santa Cruz was described as having barely 120 inhabitants who survived on marine foods and from hunting cattle, pigs, and to a lesser extent, goats. Cultivated crops included cucumber, taro (*papa china*), cassava, plantain, watermelon, melon, banana, orange, papaya, tobacco, sugarcane, lemon grass, peppermint, spearmint, otoy, avocado, star apple, cashew, mombin (*Spondia*, hog plum), fig, coconut, and date (Lundh 2006a).

46. Like most settlers in the islands, they relied on wild feral resources while cultivated crops were growing. Their "native part" of the garden included bananas, papayas, oranges, coconuts, guavas, lemons, pineapple, avocados, and yams. Cotton was provided on bushes planted by their predecessors. Their vegetable garden included beans, tomatoes, cabbage, peas, beets, potatoes, radish, cauliflower, onion, celery, spinach, sugarcane, peanuts, almonds, hazelnuts, peach, breadfruit, and avocado. Gardens had to be guarded at night from marauding animals, particularly asses and hogs. During the rainy season, they were plagued by stinging red and black ants, and nocturnal hordes of moths and butterflies. Friedo was infested with rats, which were controlled by Dora's cats (Ritter 2015; Strauch 1936).

47. Ritter and Strauch's exploits were regularly featured in the German press. Ritter (2015) is a compilation of news articles originally published in *Atlantic Monthly* 148 in

1931. Not long after their arrival, five young Germans appeared with monkeys, parrots, dogs, and rabbits, as well as a donkey that was gifted to Dora at the end of their brief stay (Ritter 2015; Strauch 1936).

48. While visiting on their yacht *Yankee* in 1933, Capt. and Mrs. Irving Johnson describe the Baroness's homestead along with thriving "doves, rabbits, turkeys and chickens" (Epler 2013: ch. 10).

49. On San Cristóbal they purchased many seeds and plants, including sugarcane, yuca, bananas, coffee, otoy, two kinds of sweet potatoes (camote), pineapple, pumpkins, mangoes, pawpaws (papaya), and a cock and two hens. They brought with them a Christmas tree from Germany which they uprooted and replanted every holiday season. On his visit to Floreana, the Wittmers were proud to show Ecuadorian Ponce-Enriques an "agricultural display including all kinds of fruits and crops, tropical, subtropical and European, that grew on the island: bananas, pineapples, mangos, figs, ciruela plums, avocados, coffee, melons, lemons, oranges, guavas, cherimoyas . . . sugar cane, potatoes, peppers, aubergines, maize and various European vegetables, also roses, hibiscus, bougainvillea, dahlias, and lilies" (Wittmer 1961: 225).

50. "Given meat, water, and salt one can stay alive and possibly happy. Our seeds would grow into vegetables and fruits—even flowers—in due time. In a place where no one else had ever broken ground, we wanted to experience, at firsthand, what it is to bring into being a way of life from the very beginning" (Conway and Conway 1947: 37). The Conways reported no cows on the islands but "hogs by the hundreds" which had to be hunted with guns and dogs, so they ate mainly goats (Conway and Conway 1947: 37). A German had brought ocelots to the island, hoping to breed tigers for sale to visiting American millionaires. Soldiers were sent to Santiago to shoot them as it was illegal to import carnivores to Galápagos "lest they escape to stock the islands and kill off the useful animals" (Conway and Conway 1947: 115).

51. His heirs built a road to the crater after having secured a new contract for salt extraction during the 1960s, but their business collapsed (Lundh 2006a).

52. Through a wooden gate framed by bougainvillea lay a path bordered by scarlet geraniums with a grove of bananas and coffee plants to the side. A downhill path led to the garden with red and green cabbages, tomatoes, cucumbers, carrots, beets, celery, kohlrabi, and beans. A few hand-tilled and manured acres were planted in potatoes, yuca, corn, sugarcane, camotes, beans, and pineapples (Conway and Conway 1947: 154–160). They also cogently point out that the bull manure used in gardening contained guava seeds, all of which sprouted successfully (Conway and Conway 1947: 269).

53. Flower seeds imported from Guayaquil included nasturtium, morning glory, sunflower, balsam, zinnia, geraniums, bougainvillea, hibiscus, rose, coral vine, and "Guayaquil flamboyants." Governor Col. Alvear considered the Conways' labors a service to the nation by introducing foreign plants and agricultural methods to the islands. "Our garden developed into a sort of botanical park and agricultural experimental station. It became our life as well as our living—our schoolbook, our theater, our baby, and eventually our ball and chain. We had failed to measure it to our own size—the size of our four hands—and so paid the price for megalomania. We did not possess a garden; a garden possessed us" (Conway and Conway 1947: 265). Expansive gardens were particularly

problematic, as they earlier mention that Galápagos were "a place where fencing is more keenly needed than perhaps anywhere on earth" (Conway and Conway 1947: 153).

54. Woram (2005: 289–297) mentions a series of likely related events: the 1921 visit of USS *South Dakota* to San Cristóbal, early 1930s surveys for potential submarine bases; a 1936 intelligence report by von Hagen for the US government, Franklin Roosevelt's 1938 "fishing expedition" to the islands, and the detailed 500-page US Office of Naval Intelligence "Field Monograph of Galápagos Islands" in 1941. A detailed account of Base Beta, and the historical contexts in which it had developed, can be found in Idrovo (2008).

55. The base was supplied with water from the San Cristóbal highlands via a pipeline that ran to the end of the pier in Puerto Baquerizo Moreno. Lundh (2006a) mentions that the pipeline gave life to the village as water became available and fisherman relocated from the interior to the coast.

56. The memorandum can be accessed at http://www.galapagos.to/WW2/19440209.HTM

57. Fishing in the relatively rich waters of Galápagos, more recently for lobsters, shark fins, and sea cucumbers (Schiller et al. 2014), served as another incentive driving mainland emigration.

58. Census figures are derived from the Instituto Nacional de Estadísticas y Censos http://www.inec.gob.ec, accessed January 11, 2017. Article 25 of the 1998 Special Law of Galápagos created three categories for human occupation: Residents, both permanent and temporary, and tourists, who require a transit control card. The government implemented "Zero People on Irregular Status in Galápagos" after INGALA estimated that 7,000 people were on the islands illegally in 2007. The policy imposes penalties for transgression, increases transparency in granting residency, and emphasizes the system of visitor transit cards (Neira 2016: 140).

59. Figures reflect 2009 and can be accessed in the *Plan Estrátegico de la Junta Parroquial El Progreso* (Consejo de Gobierno del Régimen Especial de Galápagos, 2011) http://www.gobiernogalapagos.gob.ec/wp-content/uploads/downloads/2013/08/Plan-Desarrollo-Progreso.pdf

60. Isabela was the site of an infamous prison insurrection in 1958, and today it attracts tourists to the Muro de las Lágrimas, a volcanic rock wall more than 100 m long and 8 m high constructed by prisoners as punishment, approximately 5 km west of Puerto Villamil,

61. Toral-Granda et al. (2017) estimate that 1,579 alien species have been introduced since 1535, 76% of which appeared within the last 50 years at the rate of approximately 30 per year. These include 21 marine invertebrates, 2 marine plants, 63 pathogens, 545 terrestrial insects, 77 terrestrial invertebrates, 821 terrestrial plants, and 50 vertebrates. The main pathways for introduction have included intentional importation, especially of plants, unintentional contaminants, especially associated with plants or soil, and as unintentional stowaways.

62. At the time of human discovery, native terrestrial mammals consisted of seven smaller rodents, three now extinct, and two species of bats (Phillips et al. 2012). Extant native mammals include three small rodent species of *Nesoryzomys* on Fernandina and Santiago, a small rice rat *Aegialomys galapagoensis* on Santa Fé, and two bats, the

widespread *Lasiurus cinereus*, and *Lasiurus borealis brachyotis* on Santa Cruz and San Cristóbal. Paleontological investigations have revealed specimens of a large thomasomyine rodent *Megaoryzomys curio* in the moist highland forests of Santa Cruz and Isabela, but its timing and cause of extinction are inconclusive (Steadman and Ray 1982)

63. Distributions are based on Phillips et al. (2012). Likely introduced after Darwin's 1835 visit and numerous by 1875, pigs were effectively eradicated on Santiago by 2000, the largest insular pig removal to date, with more than 18,800 pigs eliminated (Cruz et al. 2005; Lavoie et al. 2007).

64. Galápagos pigs prey on mammals, iguanas, lizards, snakes, turtles, tortoises, particularly their eggs and hatchlings; uproot nests and consume eggs and hatchlings of rails and petrels; and eat worms, beetle grubs, moth pupae, and snails (Chambers and Steadman 1986; Coblentz and Baber 1987; Hoeck 1984; Kastdalen 1982; Lavoie et al. 2007; Phillips et al. 2012).

65. They threaten native populations of land and marine iguanas, snakes, tortoises, penguins, shearwater, pelicans, flightless cormorants, blue-footed boobies, doves, finches, juvenile sea lions and fur seals, crustaceans, beetles, and orthopterans, in addition to introduced populations of cattle, pigs, burros, dogs, cats, and rats (Barnett and Rudd 1983; Coblentz 1978; Hoeck 1984; Phillips et al. 2012). Introduced mammalian carnivores are implicated in the localized extirpation of reptiles, birds, and native rodents (Dowler et al. 2000; Steadman 1986).

66. The black, or roof rat (*Rattus rattus*) is particularly ubiquitous, having spread via multiple invasions on different islands (Patton et al. 1975). Specimens of *Rattus* accumulated by barn owls were regularly recovered in the upper levels of paleontological deposits. Rats are implicated in the disappearance of native reptiles, birds, and rodents (Kastdalen 1982; Key and Muñoz Heredia 1994; Steadman et al. 1991). Rats prey on the eggs and nestlings of flamingos and dark-rumped petrels, and devastate tortoise eggs and hatchlings (Eckhardt 1972; Hoeck 1984).

67. Goats and donkeys can heavily deplete food sources on which tortoises rely, especially *Opuntia* cactus, their main dietary item (Carrion et al. 2007; Daly 1989; Fowler de Neira et al. 1985). Steadman (1986: 69) has also implicated the goat-induced destruction of *Opuntia* habitats in the disappearance of the Floreana mocking bird (*Mimus trifasciatus*). Native floras that provide shade and moisture for tortoises are also selectively damaged by feral herbivores (McMullen 1999: 38–39).

68. Goats are currently naturalized on three islands and have been eradicated on another 11 (Phillips et al. 2012). Goats can double their populations annually, as they breed throughout the year beginning at six months of age, with a 52% rate of successful conception producing twin births (Hoeck 1984: 237). They are known to have been used as political weapons in threats to stock islands with breeding goats. In 1959, one male and two females were released on Pinta Island by fishermen for future provisioning, and by 1971 their numbers were estimated at more than 20,000 (Coblentz 1978; Daly 1989; Eckhardt 1972; McMullen 1999: 38).

69. Grazing leads to the establishment of alien grasses, conversion to grasslands, and increased erosion along trails (Hoeck 1984). From a human perspective, opportunistic animals and plants may be termed weeds, defined as generally growing in places where

they are not wanted because of their perceived deleterious economic and/or environmental impact (Pyšek et al. 2004). Cattle grazing can also provide opportunities for animals like donkeys, which migrated into the humid highlands from dry coastal zones on Santa Cruz after cattle had destroyed undergrowth (Kastdalen 1982). Elephant and pampas grass were intentionally introduced to Santa Cruz for pasturage (Lundh 2006b), which was also associated with alternating changes in warbler and finch populations (Eckhardt 1972: 589).

70. Ecologically, these organisms can be considered keystone species, when their presence has a greater impact on the ecological community than other species because of their crucial role in maintaining both community organization and diversity (Mills et al. 1993). Novel or emerging ecosystems with new species compositions and abundances can appear directly or indirectly through human activity (Hobbs et al. 2006). Substitution habitats are created when native refugees are adopted into the novel habitat which is at least partially equivalent to the prior native habitat (Martínez-Abrain and Jiménez 2016).

71. The elimination of goats and donkeys has encouraged the return of dominant shrub species, especially *Scalesia*, and increased abundance of native invertebrates, birds, and reptiles in heavily grazed areas of Santa Fé, Santiago, and southern Isabela (Carrion et al. 2007; Desender et al. 1999; Hamann 2004; Lavoie et al. 2007).

72. The removal of cattle from Santa Cruz and San Cristóbal has been linked to the spread of highly invasive guava, as dense guava forests appeared after emerging shoots were no longer cropped by cattle (Eckhardt 1972: 586). Paleoecological analysis of a sedimentary record from the shore of El Junco crater lake suggests that after the 1970s reduction in local grazing intensity, pastures were invaded by alien guava, rose apple, and blackberry (Restrepo et al. 2012). Explosive expansion of introduced herbs, especially exotic fennel and star thistle, which opportunistically offered increased resource opportunity for feral European bees, was associated with the removal of cattle and sheep from Santa Cruz (Zavaleta et al. 2001: 467). The explosion of some predator populations in the absence, or removal, of other predators was earlier hypothesized as mesopredator release (Soulé et al. 1988). Related Galápagos examples might include linked population fluctuations among alien rats and native owls (Kastdalen 1982), alien cats, rats, and rabbits (Zavaleta et al. 2001), alien dogs and cattle (Hoeck 1984), alien cattle and rodents (Jones 2000), alien cattle and ticks (Kastdalen 1982), and native plants and alien insects (Causton et al. 2006).

73. Estimates vary; however, the majority of plant species in Galápagos are alien. They are found primarily in the humid highlands of inhabited islands, and commonly introduced as ornamentals, especially in coastal urban areas. A significant percentage are naturalized (Gardener et al. 2010; Guézou et al. 2010; Schofield 1989; Soria et al., 2002). Tye et al. (2002: table 1) list 37 invasive alien floras suspected of causing significant ecological change.

74. These include a wide range of alien plants, cottony cushion scale, fruit fly, tilapia, pigeon, rat, dog, cat, pig, donkey, and goat (e.g., Atkinson et al. 2012; Barnett and Rudd 1983; Campbell et al. 2004; Carrion et al. 2007, 2011; Cruz et al. 2005; Daly 1989; Lavoie et al. 2007; Phillips et al. 2012; Hamann 2004).

Chapter 3. Manuel J. Cobos, San Cristóbal, and the Hacienda El Progreso

1. Magnetic stratigraphy and potassium-argon dating suggest that the island emerged approximately 2.35 ± 0.03 million years ago with greatest volcanic activity from 2.3 to 0.66 ± 0.08 million years ago (Geist et al. 1986).

2. The Wiggins and Porter (1971) classification of Galápagos vegetation zones is perhaps the most popularly used scheme for describing islandwide floras; however, others have proposed differing classifications along altitudinal gradients (see Hamann 1981). It is followed here and combined with observations on local conditions specific to San Cristóbal (Colinvaux and Schofield 1976; Lawesson and Estupiñan 1987; McMullen 1999; Restrepo et al. 2012; Galápagos Species Checklist, http://darwinfoundation.org, accessed February 6, 2017).

3. The name Wreck Bay (also Bahía Wreck, Bahía Naufragio, Shipwreck Bay) appears on Aaron Arrowsmith's chart in Colnett (1798) and may refer to his 1794 description of a shipwreck (Woram 2016). With a tricky approach, the bay has been the site of nautical accidents, more recently the *Jessica*, which spilled 175,000 gallons of fuel in 2001, and *Galapaface 1*, carrying 16,000 gallons of fuel in 2014.

4. Many residents prefer to send their children to school in Puerto Baquerizo Moreno, where they consider educational opportunities to be better.

5. They include Soledad, Cerro Azul, El Socavón, Tres Palos, San Joaquín, Los Goteras, El Chino, Cerro Gato, and Cerro Verde. Varying statistics and irregular occupancy make it difficult to estimate how many people live in the scattered farms outside both towns; however, the number is probably greater than the 487 officially living in El Progreso. Figures can be accessed in the *Plan Estratégico de la Junta Parroquial El Progreso* (CGREG 2011).

6. Specific problem invasives include cattle, pigs, horses, goats, ants, fruit flies, aphids, various blackberries, elderberry, guava, passion fruit, pomarosa (*Syzygium jambos*), laurel, supirosa (*Lantana camarosa*), elephant grass, floripondia (*Brugmansia*), and cedar (CGREG 2011).

7. Latorre (2002: 5) cites 1871 baptismal records for a son and daughter through Cobos's marriage to Monroy's sister, Adelaida, shortly before his departure for Mexico; however, nothing is known of their subsequent fate. He later married Aurelia Baquerizo, who bore their daughter Josefina Cobos. A likely illegitimate son, Manuel J. Cobos, died in Cuenca from injuries suffered during a hunting accident in El Progreso (Latorre 1991: 109). His surviving son, Manuel A. Cobos, was also illegitimate. His great-grandson, Dagfin Cobos, supplies a good biography (Cobos 2000).

8. An 1864 company was formed by José Antonio Rubira and Manuel Jurada to exploit orchilla and fish on the islands, but it was eventually abandoned. In 1866, Cobos and Monroy brought their business interests to the island; however, in 1869 their concession was revoked (Webster 1904: 8).

9. This is a pattern that still resonates in the social and political fabric of contemporary Ecuador, which pits the liberalism of the economically and financially powerful lowland sector against its conservative, traditional, and politically powerful sierra counterpart. Highland social institutions were based in large land holdings (*latifundia*), with

roots in the early post-Conquest *encomiendas*, and involved primarily in economically underperforming activities like extensive cattle grazing. The local indigenous labor force was tied to landholders through labor service debt, or survived at a subsistence level on small land holdings (*minifundia*). Coastal landholders were faced with labor shortages, necessitating competition over free wage laborers or tenant farmers (*precarismo*). Here, production was dominated by tropical cash crops for export, notably cacao, and later bananas, coffee, tobacco and to a lesser extent tagua nuts and rubber. Export products manufactured in the highlands, especially textiles, were minor compared with coastal products, which could also be transported more efficiently along navigable rivers with access to the ocean. Toward the end of the nineteenth century, the coastal export economy generated most of the country's national income, as opposed to the conservative and traditional highland subsistence economy. Ecuador's economy has long been characterized by boom and bust cycles associated with focal export commodities, notably cacao, then bananas, and today Amazonian-based petroleum (Redclift 1978; Rodríguez 1985; Schodt 1987).

10. Numbers vary, as Wolf mentions 37 individuals during his 1875 visit, and an 1891 letter by the Territorial Marshal mentions 8 families and 20 workmen during Cobos's stay in Baja (Latorre 1991: 21). Others claim 12 families (Latorre 2002: 8; 2012: 66). Bognoly and Espinosa (1905: 86) name 10 individuals living in six small houses in 1870.

11. In 1862, the Lower California Colonization and Mining Company, formed by one of California's earliest real estate promoters, was granted 50 km^2 of land, which was parceled out to 200 shareholders, who for $220 apiece, were offered passage, 320 acres, and shares in mines near Ventana Bay. Colonization failed, and the grant was transferred to the Lower California Company in 1866, with promises of duty-free entry at the port in Magdalena. By 1871, deteriorating relations with disgruntled foreign settlers had escalated to the point where Mexican troops, having been sent to evict them, were withdrawn after a representative of the Lower California Company arrived with an American man-of-war (Kearney 1935).

12. In an 1874 complaint, John S. Howland claimed to have invested $6,000 to form a one-year business partnership in 1870 with Cobos as manager and general director to gather orchilla in Magdalena Bay. During that year, they shipped 1,400 tons of orchilla for a net profit of $228,000 which was kept by Cobos (Daily Evening Bulletin 1874). For his contract with the Lower California Company, it was claimed that he "imported 150 natives of Ecuador to assist in gathering the moss" (Daily Evening Bulletin 1871a). Also in 1871 the company advertised for pickers, offering up to $30 per month, free room and board, and 160 acres of land. Hundred of pickers had left New York by boat, and even French communist prisoners had been solicited (Kearney 1935).

13. Angel Cobos, who had been arrested by Mendizabal and the Lower California Company, was captive on board the *Cina Greenwood* when his brother allegedly seized the vessel at gunpoint. Asked why he was taking a vessel owned and commanded by American citizens, he allegedly answered, "because I have the power" (Daily Evening Bulletin 1871a). Along with Mexican officials and their peons, they commandeered the schooner and sailed it to Cayuco, where 40 armed soldiers boarded the ship (Daily Evening Bulletin 1871b). Later testimony by Mendizabal himself revealed that he had

previously seized the vessel from its owners, Cobos and Monroy, in lieu of payment for duties owed the government, because its value was greater than another ship, *Amelia*, that he had previously seized for entering goods into the country without proper documentation (Daily Evening Bulletin 1871c). Cobos was arrested when he arrived at San Francisco's luxurious Occidental Hotel on Thursday, Dec. 14, 1871 (Daily Evening Bulletin 1871a). He was described in court as "a tall, well-formed man, apparently some thirty years of age. He has a keen intelligent look, dresses in good taste, and has the carriage and manner of a gentleman" (Daily Evening Bulletin 1871b). He was acquitted of piracy and robbery charges which were "founded merely on a petty quarrel between two partners as to the ownership of a small schooner, paid for by one, but claimed by the other," and furthermore, the Mexican Collector of Revenue had sent out an official with 49 soldiers in order to compel a sub-collector of revenue (Mendizabal) to surrender his office (Daily Evening Bulletin 1872a). Kearney (1935: 38) suggests that by making an international affair out of the midnight seizure of the *Cina Greenwood*, the company had a pretext for claims against Mexico, which were eventually settled by allowing further collection of orchilla. Interestingly, an 1881 letter by M. M. Staples to the US Minister at Lima again raises the theft of two boats from Magdalena Bay, which he claims Cobos was running out of Galápagos at that time. He identifies the *Sue Greenwood*, now the *Angela Cobos*, as one of the boats. He was claiming that his own boat, *Laura*, had been seized by Cobos and his men (Woram 2005: 178; letter available at Woram 2016).

14. Leadership by traditional, church-based, sierra land–owning elites was giving way to political domination by coastal banking and commercial interests, and by 1884 a progressive era and an expanding coastal export economy were taking control (Pineo 1996; Rodríguez 1985; Schodt 1987).

15. Mann (1909: 28) was likely inaccurate when he claimed that the hacienda had been formed in 1879 "on a small croft, occupied by a negro." All earlier estimates suggest a much greater population. Also, Mann goes on to mention that the giant tortoises no longer existed due to over-exploitation, and that cattle, presumably wild cattle, were scarce during his 1907 visit. Whalers, settlers, and dogs had severely depleted tortoise populations to near extinction nearly two decades earlier on the island (Grant and Estes 2009: 85).

16. An 1886 list of island products by the Territorial Marshal lists abundant sea lions, tortoises, marine turtles, bacalao (sea bass or grouper), lobster, and wild cattle, horses, mules, and goats. Domesticated animals included cattle, horses, mules, donkeys, pigs, sheep, and goats. Cultivated plants included sugarcane, coffee, maize, potatoes, sweet potatoes, yuca, beans, pumpkins, pineapples, passion fruit, mangos, oranges, figs, ice cream bean, papayas, cherimoya or custard apples, cherries, melons, watermelons, avocados, grapes, guavas, and plantains. Orchilla and calaguala, a fern used for medicinal extract, were considered abundant, as were palo santo and paw paw resins, and matazarno, guava, and algarrobo wood. Oil was obtained from castor beans, sea lions, tortoises, marine turtles, and iguanas. Minerals included lime, salt, and sulfur (Guevara 2015: 59).

17. By 1907, Mann (1909: 28) wrote: "Orchil is to be found, but the price is too low to warrant its collection."

18. In Ecuador, 1 cuadra = 100 varas, or approximately 84 m. Eighty, presumably square, cuadras would equal 67.2 ha. In Spanish America, 1 cuadra measured approximately 10,000 m^2 (1 ha); in Argentina during the nineteenth century, 1 square cuadra was equivalent to 16,874 m^2 (1.6874 ha) (Amaral 2002). In Ecuador today, 1 cuadra is equivalent to 6,987 m^2 (0.69873 ha). At this time, different systems were popular for irrigated cane fields, or cane pieces (Galloway 1989: 90), often including rectangular beds up to 35 ft wide running parallel or perpendicular to the main dam and irrigation trench, with smaller irrigation/drainage canals or ditches crossing the beds (Deerr 1911: 107–113).

19. Jaramillo included the following: secretary, police inspector, school teacher, light house keeper, national guard chief, four officers and 50 soldiers, Cobos, his engineer, bookkeeper, administrator, clerk, three estate managers, 50 day laborers, and 55 indentured peons (Vidal Gormaz 1891: 183). Latorre (1991: 31) mentions that the soldiers were conscripted National Guard whose presence benefited the hacienda owner. Their arms and ammunition were under the control of the commander, Cobos, and the officers comprised his confidants.

20. During the 1880s and 1890s, Galápagos was serially visited, often with calls on El Progreso, by various ships and expeditions from England, Italy, Chile, and the United States. They included: *Triumph* (1880); *Albatross* and the US Commission of Fish and Fisheries (1880, 1888, 1891); *Vettor Pisani* (1884); *Chacabuco* (1887); Clark University (1889, 1891); the Webster-Harris Expedition (1898); and the Hopkins-Stanford Galápagos Expedition (1899) (Agassiz 1892, 1913; Baur 1891; Latorre 1991).

21. Each boiler produced 150 *caballos de vapor* (Martínez 1915: 47). One unit of horsepower is equivalent to 745.7 W (150 hp = 111,855 W). The majority of power generated in a sugar plantation is consumed in the sugar mill, where juice is extracted from the cut and shredded cane by multiple passes through a set of rollers or mill trains. Limewater is added to the liquor to reduce acidity, and impurities are precipitated in a clarification tank, after which the syrup is concentrated in multi-effect evaporators, crystallized in a three-stage vacuum boiler, and centrifuged. A great amount of energy is required for heating and operating the mill train, strainers, vacuum pumps, centrifuges, and crystallizers (Deerr 1911).

22. A triple-effect vacuum evaporator can be seen in the photograph (G71.5, Rollo and Ida Beck Collection, Box 40, MSS.036) taken by Rollo Beck, probably in 1905, during the California Academy of Science Expedition that visited San Cristóbal four times between 1905 and 1906 (Slevin 1931). This was an improvement on the "Jamaica Train" method of concentrating sugar through evaporation in successively smaller heated pans. The triple effect vacuum evaporator was based on the mid-nineteenth-century innovations of Norbert Rilleux, as it concentrated sugar by passing cane liquor through three vessels, each with an increasingly higher vacuum and lower temperature in order to achieve evaporation. Not only did the evaporator reduce the amount of required fuel through steam recirculation and evaporation at lower temperatures, but it also protected the sugar product from destruction at higher temperatures. When the concentrate began to crystallize (the "strike point"), it could be further filtered and evaporated to begin crystallization, and then centrifuged (Wallis-Tayler 1895: 127). Mann (1909: 31–32) describes the process in detail. After the cooled molasses was run through a centrifuge,

consisting of a revolving circular screen, the retained granulated sugar was bleached, and the skimming and expelled syrup were diluted with water and fermented into alcohol and white rum.

23. A floor plan accompanying the official 1904 visit (Webster 1904: 152) shows a larger hall, kitchen, dispatch office, bedroom, and staircase.

24. Croplands included 365 cuadras (255 ha) of sugarcane fields in three areas, Porvenir del Norte (88 cuadras), Porvenir del Sur (65 cuadras), and Lastre (32 cuadras); 172 cuadras (120 ha) of coffee in Cafetal del Sur; and 85 cuadras (60 ha), presumably of yuca, in Yucal del Sur (65 cuadras) and Yucal de Norte (20 cuadras) (Bognoly and Espinosa 1905: 166). In his 1907 visit, Mann (1909: 29) estimated more than 10,000 acres (4,048 ha) of pasture, and an earlier document claims five great pastures of 600 cuadras each (Webster 1904: 10).

25. However, Mann (1909: 32) also claimed that 1 acre of cane produced 2,500 lbs of sugar. Using this figure, and reckoning in British or metric tonnes (2,204.6 lbs), then if Bognoly and Espinosa's (1905: 166) estimations of cane fields are accurate, and all were under production at the same time, maximum annual production could have exceeded nearly twice (855.75 tonnes) his annual estimate. For comparison, these yields were on the order of three to five times less than highly productive irrigated fields in Hawaii between 1895 to 1910 (Deere 1911). An 1889 document by Jaramillo is more enthusiastic, estimating El Progreso's production at between two and three tons of high-grade sugar per hectare, better than the best production on the mainland (Latorre 1991: 40). One source estimates, presumably annual, production of around 20,000 quintales of sugar and 5,000–6,000 bottles of cane alcohol (aguardiente) (Webster 1904: 10).

26. Invited for lunch and a hunt in the highlands on November 17, 1905, Martínez commented on the cost and arduous effort required to build a road through the rock along its route. A 1904 account mentions that the 5-mile road was in great part paved with stones at a cost of 8,000 sucres (Webster 1904: 10). Agassiz (1913: 257), during his 1891 rainy season visit to El Progreso, complained about walking for an hour in knee-deep mud and broiling sun on an impassable road. In 2015 improvements were completed on a highway paved with material mined from the quarry by the airport, which begins at the eastern edge of Puerto Baquerizo Moreno and terminates at Puerto Chino.

27. El Progreso employed a state-of-the-art Decauville rail system, which began production in France for sugar beet harvests. Earliest versions consisted of small .4 m gauge track in 5.0 m units weighing 47 kg apiece, such that each unit of attached rails, sleepers, and plates could be easily carried by one man standing in the middle with one rail in each hand. The tracks, which did not sink into the mud, supported small metal cars or "porteurs," each weighing only 47 kg (Ware 1880: 218). At a reported cost of 35,000 sucres, the mobile track system was described as 5 miles and 500 m long and extending in all directions (Webster 1904: 10).

28. Another estimate corroborates 100,000 coffee plants, which had expanded in production from 600 to 2,000 quintales (Webster 2004: 9–10). Cacao, the major crop and lifeline of Ecuador's early economy, either could not grow on the island (Vidal Gormaz 1891: 186) or could grow but would not fruit Martínez (1915: 31). Farmers claim that it is grown today in the highlands, but in small quantities.

29. Estimated numbers of cattle on the island fluctuate wildly throughout El Progreso's history. For example, during his 1891 visit, Agassiz (1913: 256) estimated 20,000 head. Latorre (1991: 39) places little confidence in the fluctuating range of estimates (10,000 to 14,000) given by the Territorial Marshal, noting that at least they suggest a great abundance. It was mentioned that Cobos had planned to domesticate San Cristóbal cattle at some later date by crossing with introduced breeds (Webster 1904: 11).

30. Despite protestations by the Villamil family, as early as 1891, Cobos had constructed a 7-mile-long road from Black Beach into the island's interior to harvest wild products on Floreana. Because of predation from the many wild dogs, it is estimated that only 500 cattle were on the island at this time (Latorre 1991: 47).

31. It was estimated in 1904 that 2,000 wild burros were on the island, as well as 100 mares crossed with Peruvian horses. Pigs were being raised on a small scale (Webster 1904: 11).

32. The amount of lime in Puerto Chico was estimated at 20,000 quintales. Cobos had a building in Puerto Chico to preserve and salt fish, and he was working to have the government declare salt from Santa Elena (a government monopoly) free for use in Galápagos (Webster 1904: 10–11).

33. Both are variously referred to as a sloop (*balandra*) or pilot's boat (*pailebote*), a smaller double-masted schooner (Latorre 2011: 101). Mention is made of seven boats in all, including the 80 ton *Feliz Porvenir* and 20 ton *Josefina Cobos* (Webster 1904: 11). It is unknown whether either boat is the *Cina Greenwood* or *Laura*.

34. Martínez (1915: 39–40) reckons that a quarter of the population was composed of colonists who had lived on the island for some time. Most of the rest came from all provinces of Ecuador, especially the highlands around Ambato, as well as hill folk (*montuvios*), a respectable number of Peruvians and Colombians, various negroes from the British West Indies, a Frenchman, a Briton, an Italian, two Mexicans, two Chileans, a Hindu, and a Portoviense. An earlier document, published immediately after Cobos's death, estimates 400 workers, but only 200 hacienda employees; the rest consisted of government administrators and colonists (Webster 1904: 9).

35. Bognoly and Espinosa (1905: 100–101) illustrate examples of El Progreso money, including: 1 sucre and 50 centavo notes labeled "Ingenio Progreso-Chatam" with the signature of "Cashier A. M. Reed" on the smaller note; 100 centavo copper coins with "Ingenio-Progreso" and "Chatam" on one face and "Manuel J. Cobos" on the obverse; and a 20 centavo "caucho" (vulcanite or bakelite) coin with "Anacarsis Medina" and "Chanduy" on one face. Other variations can be found in: Seppa, Dale (2016) *Ecuadorian Coins: An Annotated Checklist*. Dale Seppa, Virginia, MN.

36. One colonial Spanish real was equal to 0.05 peso duro, or a little more than 1 g of silver. Latorre (1991: 64) points out that this wage was insufficient for survival, considering that 1 pound of meat sold in the company store for half the daily salary. He estimates that the annual salary of a day laborer could not have exceeded 50 sucres; however, Martínez (1915: 40) mentions that peons were paid a daily rate of 1 sucre on a weekly basis.

37. Martínez (1915: 40) also tells us that the majority of highland laborers ate in the hacienda, depositing the remainder of their wages with an Administrator. They would

regularly remit funds to Ambato for their families, pay off debts, or buy land, which was their supreme ambition.

38. The dance hall was described as a 19 m by 5 m zinc-roofed hut with a floor of Roman cement. Everyone was required to pay 70 centavos whether or not they wished to partake in activities. On certain occasions, when a new employee arrived or a birthday was to be celebrated, work did not end until sunset, so they danced in the light of a burning lamp that dispensed thick, black smoke to music provided by a band composed of guitar, violin, and accordion. Along one wall, cane liquor (*aguardiente*) was supplied, and we are told that these social events were in truth always theaters of bloody scenes and grand orgies (Bognoly and Espinosa 1905: 102).

39. An April 1904 census counts 145 men and 36 women between the ages of 18 and 40 (Guevara 2015: 58). If we add the 78 men and 8 women captured from the *Libertad* in February 1904 (Bognoly and Espinosa 1905: 92), the figures increase to 225 men and 44 women (16%). Fluctuating ratios of men to women at El Progreso are provided for the following years: 1887: 140:20, 107:60; 1889: 123:44, 188:53, 213:54. The proportion of married partners never exceeded 40% and was usually far less (Bognoly and Espinosa 1905: 89; Guevara 2015: tables 4 and 5; Latorre 1991: 59). Bognoly and Espinosa (1905: 104) mention that some women had been married up to seven times.

40. Perhaps the most famous example is that of Camilo Casanova, a rebellious worker who threatened Cobos's life after having been lashed 400 times for attempted murder. He was marooned on Santa Cruz with a canteen of water, blunt knife, machete, small hatchet, 18 boxes of matches, and a few clothes. Surviving at first on raw fish, iguanas, and turtle blood, he eventually left the beach for the interior, where he found water, a plantain grove, yuca, sweet potatoes, chickens, oranges, and mamey or zapote, which had been planted years earlier by other exiles. He survived for three and a half years hunting birds, fishing sea bass, and growing fruit. Passing ships would leave him supplies but refused to liberate him because of a multilingual sign that had been posted on the island: "Please do not take this man away because he is twenty times criminal." He was eventually rescued three months after Cobos's death and reunited with his wife, who had been told by authorities that he had died and left her in debt (Bognoly and Espinosa 1905: 106–110).

41. Both bodies were disinterred and autopsied on February 28, 1904, in the dance hall. Reina's was found to have a bullet entrance wound at the base of the neck above the left shoulder, causing injury to the cervical arteries and left lung before exiting the middle of the back. A stab wound to the belly also caused protrusion of the intestines. Cobos was found to have four bullet wounds, one above the left nipple, one in the left leg, and two in the back. Also, the left leg was completely fractured, presumably from the fall, and the lower jaw was broken by postmortem blows. Both bodies were subsequently removed for burial in Guayaquil (Webster 1904: 132–134).

42. In a letter to his friend Leonardo Plaza, the president of the Republic, newly installed Territorial Marshal Juan José Pino wrote that life for island residents was poorer than in Siberia, and that Cobos had declared "God in Heaven, and I in Chatham." Cobos, he said, had been worse than Roman tyrants, yet the real culprit was poor government.

In a later assessment, he lamented the lack of a system of colonization except for monopolies that developed into absolutism, further noting the absence of welfare on islands where "the only certainty here is the grave" (Latorre 2002: 57–58).

43. It is worthwhile here to quote Alexander Mann's (1909: 35–36) description of Cobos at length: "I was acquainted with Cobos, and found him a most pleasant conversationalist, fairly educated, practical, and even humorous, and honourable in his business engagements, reminding me of Byron's pirate, 'the mildest-mannered man that ever scuttled a ship or cut a throat.' He had lived for twenty-five years, an autocrat among a criminal community, carrying his life at his revolver belt, and ruling with a rod of iron. He had redeeming points in his character, but in that settlement the field for humanity was circumscribed, in a civilised country, restrained by efficient laws, Manuel Cobos would have been an able pioneer of progress and a respected citizen."

44. The sucre became the national currency in 1884. Mann (1909: 34) estimated the hacienda's worth at £30,000, which (using a 1904 exchange rate of 4.87) would have been equivalent to US$146,000. Its net worth (project value and economy cost) in 2016 could have been as high as US$31 million. https://www.measuringworth.com/ (accessed May 8, 2017).

45. In that same year, Galápagos suffered its first petroleum spill, as 23 tons of fuel were accidently pumped out of the small steamship *Cristóbal Colón* into Wreck Bay during the night (Latorre 2011: 121).

46. Expedition engineer Ludwig Naess exclaimed, "A factory such as this one I have never seen or heard of in all my life. The building was made of bamboo and rusty sheets of corrugated iron. The machines were old and highly unsanitary. The pipes through which the molasses flowed were not screwed together as we do it, but were joined by crimping the end of one pipe and forcing it into the end of another pipe. As a result, much of the molasses was lost. Repeatedly, the machines overheated, and the poor foreman was scolded by the owner" (Hoff 1985).

47. In 1924 Cobos eluded death at the hands of revolting workers who pillaged his reserves of alcohol, looted the commissary, destroyed machinery, set fire to cane fields, and burned documents. Cobos had escaped on horseback, but the torture of a trusted servant revealed his hiding place. Facing death, Cobos opened the hacienda safe, which contained no money. His enraged assailants forced Cobos to compose letters freeing them from their criminal sentences. The house was subsequently burned, after which they escaped on the *Manuel J. Cobos* and were eventually captured in Esmeraldas (Egnal 2015).

48. Interestingly, he mentions leaving a "little brown honey bear" (kinkajou, *Potos flavus*) with Karin Cobos; he had acquired it a month earlier in San Miguel Bay as a pet (Robinson 1932: 60).

49. After 1935 motions by various Anglo-American expeditions, the sanctuary was created "to promote the development of the sciences and assist in every manner possible to realize such cultural progress" and prevent further species loss due to "depredations of unscrupulous travelers and tourists" (Ackermann 2014).

50. Another visit in 1938 by Paulette de Rendón describes the new hacienda house

constructed by Tous in the exact location of the Cobos house, ruins of which partially survive today. On the first floor was the hacienda store, the only one on the island, "where all that this miserable population of peons may need is sold at a dear price" (Latorre 2002: 72–73).

51. Territorial Marshal Col. Meneses wrote in 1942, "All of the officials and workers have not been paid since January. There are neither supplies nor clothes, because there is no money circulating. Previously, all these articles were provided by the hacienda, but now the hacienda exists only in name, because its owners, still without title, don't order more. They affirm that they are blacklisted" (Latorre 2002: 73).

52. Arthur Eichler sailed to San Cristóbal in 1950 with an Italian who told him that he was on his way to liquidate a canned fish business in which he had lost every penny. Visiting at a time when the island had 802 residents, Eichler also encountered Karin Cobos, who had never left the island (Eichler 1955). In the following year, William Crealock visited El Progreso, referring to it as "an unattractive, unimaginative little settlement" of huts housing a few hundred inhabitants and that "there was no attempt made to beautify it in any way" (Crealock 1955: 22).

53. "together with science we move forward."

54. None of the first boatload of colonists had any experience fishing for lobster, intended to be frozen and returned for sale in the United States for $40,000 profit. Neither lobster nor tuna were profitable at the time for local fishermen. The freezer in town, which had been purchased for $30,000, and the refrigeration unit on the *Alert* were damaged. In 1936, the Galápagos Islands had been declared state property, thus legal ownership of land was in dispute. As for their gardens, one colonist declared, "Pigs, burros, and chickens ate everything we planted. So we built a fence. . . . No sooner than the second crops began to shape up nicely, the rains came. For weeks on end the sun did not appear on the hill. Our garden died" (Faris et al. 1964: 52).

Chapter 4. Island Production for Global and Local Consumption

1. *Ingenio* (genius, creative ability) and more specifically *ingenio de azúcar* refer to a sugar mill, often including the entire sugar operation; whereas, *trapiche* (sugar mill) identifies the grinding machinery and the mill itself (Moreno Fraginals 1976: 82).

2. The northern coast of Peru was an early site for sugarcane production, much of which was directed via trade to markets in Chile. During the latter half of the nineteenth century, Peruvian production began to supply the growing demand for sugar in Western Europe and the United States, while maintaining a virtual monopoly over expanding Chilean markets. The latter came to an end in 1879 during the War of the Pacific, which pitted Peru against allied Chile and Bolivia. Sugar exports to the United Kingdom, which began in the 1870s, eventually came to an end at the turn of the century as a result of oversupply, when Peruvian sugar was redirected to Chilean and U.S. markets (Gonzales 1985).

3. The *casa de vivienda*, housing the hacienda's owner or administrator, was regularly placed on the highest ground overlooking the entire *ingenio* (Hazard 1871: 351; Tezanos Toral 2015: 69).

4. Characteristics of this feature, including its form, construction, and materials are similar to ovens used on the mainland in the sixteenth through eighteenth centuries (Bolaños et al. 2002).

5. Papers were filed by the Junta Parroquial in July 2017 to recognize the hacienda house site and the nearby tomb, as national historic sites. The tomb site, which has also received touristic amendments, including a thatched shelter over a commemorative cement grave painted white with surrounding benches and plantings, is not the final resting place of Cobos. This is likely the area in which the bodies of Cobos and Reina were hastily buried after their assassinations, and it is believed to be the spot where five workers were shot in 1880 by order of Cobos. Both bodies were exhumed for autopsy and reinterred in Guayaquil (Webster 1904). Subsequently, papers were filed in 2018 to include the sugar mill area as national patrimony.

6. In 2014, we received permission from a family member who resides in El Progreso to enter the properties, which were for sale at the time. We made two inspections of the area and exposed a small profile for phytolith sampling, but on our third visit in the company of the parroquia president, we were vigorously asked to leave by another family member. The property was sold in the following year to a coastal businessman who turned his parcel into a paint ball emporium and also strenuously denied access. In 2017, the property was again listed for sale, when we were kindly allowed to enter the properties. A quick perusal was undertaken, and we returned in 2018 to map the area. In their earlier INPCE study, Bolaños et al. (2002) were also denied access by the property owners; their limited observations were based on information supplied by a worker contracted to the owners.

7. The location of the midden was fortuitously revealed in 2014 when the carpenter showed one of us various artifacts that he had found on his property. When we returned to inspect the exposed profile, he mentioned that it was his wife's wish to expand the existing courtyard by cutting into the back wall for an intended outdoor café from which she could sell her baked goods. The carpenter had not found the time to undertake this project; however, we suggested that we could excavate it for him while removing a portion of the midden. His enthusiasm was piqued upon finding out that we would perform this task at no charge.

8. A floor plan of the hacienda house (Webster 1904: 137) shows an accounting office close to the staircase leading to the second floor. Latorre (2002: 36) maintains that immediately adjacent to it was a room that served as the jail. This room, located next to Cobos's office, is labeled in the floor plan as "Ayudante" or assistant. There is no clear reference to this building ever having served as a jail. We have also heard of it referred to as "El Comiseriato," which corroborates Bolaños et al. (2002), who maintain that it had served as a town grocery store until relatively recently.

9. In the field, soil samples were processed using bucket flotation in approximately 20 L of clean water. A sieve of 250 μm was used to collect the light fraction, which was then transferred to cloth bags. Light fractions were sieved in the laboratory in order to separate components. Wood charcoal specimens were dried to room temperature and then placed in small clean plastic bags. Fourteen bags of charcoal were collected during excavation and a subsample of 288 fragments were analyzed. A maximum of 20 charcoal

fragments were randomly collected from each bag. In addition, two fragments of wood were extracted from a tree trunk that remains underwater close to the modern city of Puerto Baquerizo Moreno. It appears to be one of the remaining original supports from the nineteenth-century pier of the Hacienda (Astudillo 2017: 70).

10. In all cases, the soil matrix was troweled sequentially from lower to upper levels into sterilized 15 ml plastic test tubes. Sample units were spaced 5 cm apart and visible horizon boundaries were avoided. Each sample contained a maximum volume of 10 ml. The sampling tool was washed after each extraction with distilled water to avoid cross-contamination. Altogether, 31 samples were collected from the archaeological contexts; 18 from the Carpintero location and 13 from the sugar mill. A total of 52 samples, 13 from each column, were collected, including topsoil, which usually contains the highest phytolith concentration (Fishkis et al. 2009). Altitude, extant vegetation, and soil type and color were noted. Reference specimens of local trees and grasses were collected for phytolith extraction (for expanded discussion, see Astudillo 2017, 2018a, 2018b, 2018c).

11. The nineteenth-century Cuban *ingenio* included a "prominent, large and central" sugar factory composed of three important buildings. The Milling House or Casa de Ingenio was often no more than a shed to protect the steam engine and cane press. The Boiling House or Casa de Calderas received the extracted juice from the mill for defecation, filtration, and evaporation. The Purging House or Casa de Purga was generally the largest building, often two stories high with an open side to receive boxes of sugar for drying on the upper floor while draining molasses through to a lower floor, followed by drying in racks (Tezanos Toral 2015: 80–86).

12. Many artifacts were removed to new locations, often for decoration, like the spur wheels, pinions, and press rollers in El Rondel and the entrance to El Ceibo. Many of the more conspicuous artifacts of any value were believed to have been transported to the naval base for curation; however, none were acknowledged by base officials nor seen during frequent visits. At times we searched for specific items, and we often located them accidentally or through word of mouth.

13. Where viable, water power had been the preferred source for driving cane mills; otherwise, wind and animal power were utilized. The adoption of steam power for sugarcane processing had been described as early as 1768 by John Stewart, and first successfully applied in Cuba in 1797. By the mid-nineteenth century, steam power was widely used in the Western Hemisphere (Deerr and Brooks 1940). The first steam-powered facility in Peru may have been in use during the 1860s at Facalá in the Chicama valley (Gonzales 1985: 54).

14. We surveyed the surrounding countryside for preserved earthen canals that might have been associated with water transport and field irrigation. One such feature is located at El Trino, a farm southeast of town. The western end of the feature terminates in a large depressed meadow which could conceivably have been used for water retention. It is, however, difficult to associate the canal with the historic hacienda reliably.

15. In their earlier inventory, Bolaños et al. (2002) illustrate an approximately 1 × 0.5 m metal door frame used to control water flow in canals. They also mention semisubterranean and subterranean tubes connecting the hacienda house and *ingenio* with water reservoirs. Fragments of pipes litter the area.

16. Each opening was connected by a horizontal tube to its counterpart on the opposite end. Heat was conducted from the externally fired dutch oven through each tube via a flue on the back end of the combustion chamber. The multiple tubes created a large surface area for heating the surrounding water within the sealed boiler (Gutiérrez Ortiz 2011). At least 114 openings can be counted on one intact boiler end. Deerr (1911: 404) calculated that a 7 foot diameter boiler with 120 tubes can supply 1,800 ft^2 of heating surface.

17. Originally founded in 1857 as the Buffalo Agricultural Machine Works, the Squier brothers and partner John Valentine established a foundry to manufacture agricultural machines and implements (Anonymous 1866: 66), which in 1884 became the George L. Squier Manufacturing Company after the death of Henry Squier in 1882 (Hubbell 1893).

18. An improved McOnie Harvey & Co. triple effect vacuum evaporator from the 1893 World's Columbian Exposition in Chicago is shown connected to a horizontal steam engine, and closely matches in appearance the El Progreso apparatus (Anonymous 1893: 228).

19. Either side of the basket is inscribed in flowing script with "Jose Mª Vallejo/ Dbre 30 de 1908" on one face, and "Administrador / [B?] R. chabarria/ Enero 1 de 1909" on the other. We suggest that this might commemorate a succession of administrators over the New Year.

20. The fragment of a third wheel is currently in the archaeological collection at El Progreso.

21. Mann (1909: 54, and see Wolf 1887: 21) mentioned how the immense herds of Isabella Island were tamed by sending out domestic decoys that returned to their pens with wild cattle, which were subsequently starved to the point of docility. Alternatively, domesticated cows could be released into open pastures to mate with wild bulls. Webster's (1904: 11) comment that Cobos had planned to breed local stock with Durban cattle is probably a confused reference to Durham cattle. The Durham or Shorthorn was originally developed for dual use in northeastern England during the late eighteenth century, and was later used in South America to upgrade local stock (Rouse 1970: 312–315).

22. A detailed account of the excavated archaeofaunal assemblage (Stahl 2017) will become available online through the Alexandria Archive Institute/Open Context, and was included in our report on file with the Instituto Nacional de Patrimonio Cultural in Ecuador, available upon request.

23. Mann (1909: 32) describes "lots of nice domesticated cattle," larger than those he knew from home, "with very clear skins." Martínez (1915: 45) mentions cattle of "diverse colors" that on Isabela were "beautiful and corpulent" (1915: 94). Before the later nineteenth-century introduction of foreign breeding stock, all Criollo cattle were descended from the small population of cattle originally introduced by Columbus. The earliest cattle in Ecuador were shipped from Panamá, where breeding herds based on Caribbean stock were established in 1510 (Rouse 1977). Standard measurements taken on lower weight-bearing bones in the assemblage suggest that El Progreso cattle are similar in size to smaller, northern cattle from later periods (Reitz and Ruff 1994). Limited evidence from metapodial elements in the midden also suggests that at least a few of the animals that contributed to the sample were small females (Higham and Message 1970; Grigson 1982). Using criteria developed by Matolcsi (1970), one of the cows in the midden may

have measured 1,266.9 mm (50 inches) at withers (highest body height, usually taken at the hip), 1,599.6 mm (63.3 inches) in body length, and weighed a maximum of 143.4 kg (315 lbs). Breed assignation based on preserved bone material is difficult. Comparing a variety of specific weight-bearing bones in the assemblage with those from European breeds (Higham 1969; Noddle 1973) suggests that the El Progreso cattle were very small. Withers height compares well with coastal Jaspeado Manabita cattle; however, they weighed much less (Cevallos-Falquez et al. 2016; Porter et al. 2016: 173). Having been introduced no earlier than the 1830s, it is doubtful that this can be attributed to island dwarfism; however, it is possible that the original founder population was selected for smaller size to accommodate their shipment nearly 1,000 km to the island.

24. *Bos* contributes the highest number of identified specimens (NISP) and minimum number of individuals (MNI) in the Carpintero sample: NISP = 1,261, MNI = 27 based on the right calcaneum, wt = 34,304 g. Age at death estimates are based on published schedules of bone fusion and dental eruption (Grigson 1982; Habermehl 1961; Silver 1970) as well as arbitrary approximations of tooth wear (Grant 1982).

25. The correlations between differential specimen survivorship (Lyman 1994: 234), bone mineral density (Kreutzer 1992), and within-bone nutrients (Brink 1997) were consistently weak and statistically insignificant (Stahl 2017).

26. *Capra* specimens are the second most common identified mammal in the midden: NISP = 221, MNI = 6 based on the left distal Humerus, wt = 1,485 g. An estimate from measurements of two complete calcanea yield withers heights of 96.9 cm (38 inches) and 60.4 cm (23.8 inches). The smaller estimate may be of Native Creole goat weighing approximately 26.6 kg. The larger estimate is possibly a Nubian, one of the tallest and largest goat breeds, weighing 110–140 kg. A complete, 106 mm long, backward-curving horn was recovered from the midden. Both Nubian males and females sprout horns, and young males reportedly can grow 200–250 mm long horns in their first year. (Kysely 2016; Porter et al. 2016: 357–358; Teichert 1975; Vargas et al. 2007).

27. Pooled *Capra* and unidentified medium-sized artiodactyl specimens were used to estimate age at death based upon epiphyseal fusion (Habermehl 1961; Noddle 1974), and dental eruption (Bullock and Rackham 1982: 77) and wear (Grant 1982).

28. *Sus scrofa* (NISP = 26, MNI = 3, based on right maxilla, wt = 161) age at death estimates are based on bone fusion, dental eruption, and tooth wear (Bull and Payne 1982: 66; Grant 1982; Habermehl 1961: 127).

29. Most elements of the dog (*Canis lupus familiaris*) skeleton appear to have been preserved (NISP = 66, MNI = 3, wt = 179 g), and the majority (NISP = 53, 80%) were recovered from the same unit and level (U1 L.7). Age at death estimates are based on bone fusion (Habermehl 1961: 143). *Felis catus* specimens (NISP = 15, MNI = 2, based on complete right humerus, wt = 11 g), included two dentary elements with fully erupted molars, and no immature postcranial specimens (Habermehl 1961: 151).

30. Two species of *Rattus* (NISP = possibly 8, MNI = 2), including the smaller Black Rat (*R. rattus*) and larger Norway Rat (*R. norvegicus*), are found on most islands, but the Black Rat is most common and widespread. Endemic rodents are not found on San Cristóbal; where they are found in Galápagos, they tend to be smaller than, or in one case overlap in size with the smaller *R. rattus* (Tirira 2007). Almost 80% of identified

Chicken specimens (*Gallus gallus*, NISP = 38, MNI = 4, based on distal left femur, wt = 39 g) are leg and wing elements.

31. Currently, at least 25 species in 17 genera of sea basses and groupers (Serranidae NISP = 1,146, MNI = 40, based on right premaxilla, wt = 2,481 g) are recognized in Galápagos (Froese and Pawley 2017; Ruiz et al. 2011). Rough size estimations based upon complete maxillae, employing the proportional measures published for Sailfin/Galápagos Grouper or Bacalao (*Mycteroperca olfax*) published by Smith (1971: 207), indicate groupers between 51.9 and 67.6 cm in length, although Bacalao can attain total lengths up to 120 cm.

32. The majority of Cheloniidae specimens (NISP = 766, MNI = 3 based on sided limb elements, wt = 5,074 g) comprise shell (carapace/plastron) elements; therefore the MNI estimate is a very conservative estimate.

33. Recent analysis of aDNA recovered from two turtle specimens in archaeological deposits in San Francisco also confirms the exploitation of Green Turtle during the California Gold Rush (1848–1855). Eastern Pacific populations of Green Turtle declined in abundance to near extinction in the twentieth century (Conrad et al. 2018).

34. Giant Galápagos Chiton (*Chiton goodalli*, NISP = 535, wt = 1,263 g) attains maximum lengths of 11–12.5 cm, whereas the smaller Sculptured Chiton (*C. sulcatus*, NISP = 53, wt = 77 g) does not exceed 9.5 cm. All other native chitons are considerably smaller and less abundant (Smith and Ferreira 1977). The larger chitons are pried from intertidal rocks under the light of the full moon as they emerge to feed on microalgae. Although the Sculptured Chiton grows faster, Giant Chiton is preferred for its size and volume of edible meat, which is usually consumed in ceviche (Herrera et al. 2012; Murillo Posada 2010).

35. Concentrations of quadrilobate and saddle morphotypes increased in the historic contexts of the midden; neither morphotype was observed in contexts prior to human occupation. Quadrilobate phytoliths are commonly formed in maize (*Zea mays*), and saddle phytoliths were observed in leaves of sugarcane (*Saccharum officinarum*); both crops were early introductions to the Galápagos Islands.

36. Modern vegetation is represented by high concentrations of phytoliths of *Musa* sp., palms, and globular phytoliths preserved in surficial layers. These plants were also growing on the surface of the midden area when the soil samples were collected, and could be intrusive.

37. Palm phytolith morphotypes can be classified into two groups of subfamilies; however, lower resolution identification is elusive. Two possible identifications in the subfamily Arecoideae include Alexander or Solitaire Palm (*Ptychosperma elegans*) and Coconut Palm (*Cocos nocifera*). Both are found today on the island, and the latter was earlier identified by Martínez (1915: 27) as the cultivated palm on San Cristóbal.

38. A total of 289 wood charcoal fragments included: *Piscidia carthagenensis* (*n* = 176); *Scalesia pedunculata* (*n* = 27); *Psidium galapageium* (*n* = 20); *Pinus* sp. (*n* = 17); Unidentified species 2 (*n* = 12); cf. *Croton scouleri* (*n* = 11); Unidentified species 1 (*n* = 10); *Guadua angustifolia*/*Bambusa* sp. (*n* = 8); cf. *Bursera graveolens* (*n* = 6); and *Quercus* sp. (*n* = 2) (Astudillo 2017: 76; 2018a).

39. Various photographs of Hacienda El Progreso taken during the 1888 visit by the USS *Albatross* scientific expedition show large pieces of Matazarno trunks forming the structures of some plantation facilities, the main house, and parts of the workers' houses.

40. Only small fragments of charred bamboo were identified in the wood charcoal assemblage of El Progreso, suggesting its high combustion and poor preservation in the archaeological record. Two styles of local architecture can be seen in a photograph taken by the Alexander Agassiz expedition in 1891 (Image 744089, Archives of the Museum of Comparative Zoology, Ernst Mayr Library, Harvard University) of El Progreso village. A cabin made with bamboo walls and a house made with bahareque walls (clay mixed with grass or ash over a bamboo frame) both include bamboo. Its stems form the wall and balcony frames of both houses. Their rooftops appear to be made with bamboo, banana, or sugarcane leaves (Astudillo 2017: 80).

41. Arboreal vegetation was recognized by concentrations of a globular sinuate morphotype and phytoliths from the Asteraceae. Globular phytoliths are formed in most tree and shrub taxa in tropical zones (Piperno 2006), and along with ovate phytoliths are formed in leaves and stems of native taxa in the Galápagos. They were observed in the comparative material of large native trees, including *Piscidia carthagenensis*, *Psidium galapageium*, *Scalesia pedunculata*, and *Prosopis juliflora*. The phytolithic record supports the wood-charcoal record in suggesting that the local landscape prior to human occupation was covered by tall trees, possibly *Scalesia*, *Psidium*, and *Prosopis*, along with short native sedges. Concentrations of bilobate phytoliths in the historic midden suggest the early appearance of introduced grasses. In this phase, a low presence of the parallelepipedal bulliform cell and globular phytoliths suggest vegetational change from a native to anthropogenically modified landscape after deforestation. The low numbers of globular phytoliths in the midden could be evidence of deforestation, or they could be due to poor preservation, or melting of phytoliths during garbage burning. Grass is represented by large concentrations of the parallelepipedal bulliform cell, a phytolith formed in the leaves of most C3 and C4 grasses. This morphotype is also observed in the native species *Rhynchospora nervosa*, a grasslike plant member of the Cyperaceae. Other grass phytoliths such as scutiform lanceolate and elongate cells are also present in this phase. These two morphotypes were observed in native *Scleria melaleuca* Rchb. ex Schltdl. & Cham., a grasslike plant member of the Cyperaceae and in *Stenotaphrum secundatum*, a native grass (Astudillo 2018b).

42. A decrease in D/P ratios tracks change from forest to open or savanna-like habitats (Coe et al. 2013, 2014, 2015; Evett et al. 2007). Indices were obtained from phytolith concentrations following Alexandre et al. (1997) and Bremond et al. (2005). A total of 56.149 g of soil was processed for phytolith extraction. Biogenic silica averaged 6.27% of the total dry mass and 8,076 phytoliths were counted. Five main groups of phytolith morphotypes were identified: (1) panicoid phytoliths (scutiform lanceolate, bilobate, polylobate, cuneiform bulliform, and quadrilobate morphotypes), (2) elongate phytoliths (grass phytoliths), (3) globular phytoliths (arboreal phytoliths), (4) other diagnostic, and (5) non-identified morphotypes. Bilobates, polylobates, and quadrilobates are morphotypes commonly associated with panicoid grasses (Piperno and Pearsall 1998).

With an average length of 20 μm, bilobates are common in the genera *Aristida* (Arundinoideae), *Eragrostis* (Chloridoideae), and *Stipa* (Pooideae), all of which are present in the soils of San Cristóbal Island.

43. The counts of diagnostic grass phytoliths (bilobate, cuneiform bulliform cell, saddle, elongate epidermal long cells, and parallelepipedal bulliform cells) all formed in Poaceae grasses, and noticeably increase in contexts associated with human activities, especially after the onset of industrial-scale agriculture at El Progreso plantation. These phytoliths are formed in leaves of some members of the Panicoideae subfamily of grasses such as *Andropogon*, *Aristida*, *Eragrostis*, *Panicum*, and *Zea* (Mulholland 1989). Today, these genera are considered introduced and invasive taxa throughout the archipelago and dominate modern local ecology on various islands.

44. Two species of agave (*Agave americana* and *A. angustifolia* var. *marginata*) were introduced into San Cristóbal Island during colonization. Agave fences were common in the Ecuadorian highlands at least until the importation of barbed wire, which was invented in 1879. Natural fences made with trees can be observed in the historical photographs of El Progreso.

45. A high-resolution pollen record (Restrepo et al. 2012) from El Junco indicates a preclearance landscape dominated by shrubs and small trees, including co-dominant Euphorbiaceae (*Acalypha*) and Chaff Flower (*Alternanthera*), along with Croton (*Croton scouleri*), Culantrillo (*Brickellia*), Pilea (Urticaceae), Nightshades (*Iochroma ellipticum*), Scalesia (*Scalesia*), Buttonweed (*Borreria*), and Miconia (*Miconia*). Many exotic trees and larger shrubs were subsequently introduced by humans from tropical areas around the world for various uses. On San Cristóbal, these include: Ceibo (*Ceiba pentandra*), Palo Santo (*Bursera graveolens*), Papaya (*Carica papaya*), Matazarno (*Piscidia carthagenensis*), Citrus (*Citrus*), Palo Verde (*Parkinsonia aculeate*), Avocado (*Persea americana*), Spanish Cedar (*Cedrela adorata*), various Algarrobos (*Acacia, Prosopis*), Cat's Claw (*Zanthoxylum fagara*), Tamarind (*Tamarindus indica*), Chinaberry (*Melia azedarach*), Common Guava (*Psidium guajava*), Rose Apple (*Syzygium jambo*), and Coconut Palm (*Cocos nocifera*). Other islands now also support: Balsa (*Ochroma pyramidale*), Tropical Almond (*Terminalia catappa*), Inga (*Inga schimpffi*), Jaboncillo (*Sapindus saponaria*), Oleander (*Cascabela thevetica*), Butterfly Flower (*Bauhinia monandra*), Poinciana (*Delonix regia*), Flame Tree (*Erythrina velutina*), Teak (*Tectona grandis*), Tulip Tree (*Spathadea campanulata*, Malay Apple (*Syzygium malaccense*), Quinine (*Cinchona succirubra*), and Australian Pine (*Casuarina equisetifolia*) (McMullen 1999).

46. Repeat photography developed out of late nineteenth-century glaciology to become an important tool in the study of ecological and geological change in landscapes (Webb et al. 2010). The methodological procedures followed in this study are found in Taggart-Hodge (2016).

Chapter 5. Consumption and Control in the Material Culture of Hacienda El Progreso

1. In all cases these artifact counts and weights do not include the (considerable) faunal material recovered.

2. Comparable archaeological materials come from Hacienda Tabi, in the Yucatan.

Here Sweitz (2012: 131–132) found one 1807 coin, 9 Mexican coins from 1900 to 1910, and a single roughly cast metal token, marked "TABI/MECATE." A mecate is a 20 m square area of land, creating a bundle of henequen (which could be redeemed for the token). Sweitz suggests that the hacienda ran largely without coinage, but with some tokens, until it was sold in 1907, at which point the new owner may have introduced the use of national coinage for paying workers. In comparing Tabi with El Progreso, we can see a strong similarity, in that we recovered no nineteenth-century Ecuadorian official coinage, and the commission of 1904 saw the El Progreso workers' lack of knowledge of the national coinage denominations as a sign of how isolated and poorly treated they had been.

3. Unless otherwise indicated, all bottle analysis terminology is taken from the Society for Historical Archaeology bottle analysis website, at https://sha.org/bottle/

4. Square glass alcohol bottles, or "limetas" were all used for gin, as by "last quarter of 19th century" that is all they were used for (Giovannetti and Lema 2007: 95; Pedrotta and Bagaloni 2005: 186).

5. Liquor and wine bottles vary in weight, but using an average from recovered El Progreso complete bottles along with examples of similarly dated wine and liquor bottles from the SFU Archaeology collection, we get a reasonable average of 650 g for a medium-sized empty liquor or wine bottle of Cobos's era.

6. http://bardinet.fr/en/our-company (accessed June 23, 2018).

7. Examples of Siegert's Bitters bottles have been recovered archaeologically from the port of Islay, Peru (Ravines 2008: sp. 953), the nineteenth-century leprosarium at St. Thomas in the US Virgin Islands (Barton 2012: 112), the US Army post at Fort Bowie, Arizona (Herskovitz 1978: 16), and the Mascot Saloon in Skagway, Alaska (Spude 2005: 272).

8. Presumably either *chicha,* or another fermented beverage brewed by the workers themselves and then stored in used glass bottles.

9. https://sha.org/bottle/food.htm (accessed June 23, 2018).

10. https://blog.qm.qld.gov.au/2017/11/15/a-sperm-sewing-machine-oil-bottle-from-aarhus/ (accessed June 23, 2018).

11. All these white ceramic buttons were made by the Prosser process, patented in 1840. White china buttons were exported all over the world from the 1840s onward by England, France, the United States, and Germany (Sprague 2002).

12. The various censuses undertaken by territorial marshals between 1886 and 1904 included tailors and seamstresses among the occupations at El Progreso.

13. http://oldglassbottles.blogspot.com (accessed June 23, 2018).

14. The printed address on our example, at 533 Oxford St in London, suggests that the pot was manufactured between 1868 and 1909. http://www.ointmentpots.com/victorian/holloways (accessed June 23, 2018).

Chapter 6. Galápagos, San Cristóbal, El Progreso, and *Colonos* in a Changing World

1. The model, which considers strict nature preservation as its primary goal, generally with the exclusion of local resident populations, has been criticized on ethical and pragmatic grounds for decades. Particularly in developing countries, any accrued benefits

are attained at the expense of disenfranchised local populations. Traditional Ecological Knowledge, local institutions, and social organization are neglected, negating opportunities for grassroots development. Protection has been cited as costly, ineffective, and even counterproductive to the aims of conservation (see Schelhas 2010).

2. Latorre (2011) has suggested nine overlapping stages in human history in Galápagos: prehistory; Spanish discovery and early visitation; invasion by privateers and whalers between 1684 and 1864; Republican annexation in 1832; epoch of haciendas between 1860 and 1920; territorial headquarters and administrative confusion from 1884 to 1946; scientific evolution; the Ecuadorian navy from 1946 to 1979; and the establishment of a new province in 1973.

3. According to the IUCN Red List (IUCN 2018), archipelago-wide animal extinctions include three endemic rodents: Darwin's Rice Rat (*Nesoryzomys darwini*), Indefatigable Galápagos Mouse (*Nesoryzomys indefessus*), and Giant Galápagos Rat (*Megaryzomys curio*). The Vermilion Flycatcher (*Pyrocephalus dubius*), a San Cristóbal endemic, is presumed extinct, particularly through nest predation by accidentally introduced rats and the parasitic fly *Philornis downsi* (Carmi et al. 2016; Tellkamp 2017). The Galápagos Hawk (*Buteo galapagoensis*) has been locally extirpated from San Cristóbal. Recent studies on Santiago Island have documented the hawk's dietary shift after goat extirpation and the potential for problematic effects of associated vegetation regrowth (Jaramillo et al. 2016). The native Chatham Island Tortoise (*Chelonoidis nigra*) had been earlier extirpated over most of its native range on the island through hunting, and by nest trampling and predation by invasive animals. Today, various tortoise species are being successfully reintroduced throughout the islands from a breeding program located at La Galapaguera. The endemic Galápagos Amaranth (*Blutaparon rigidium*) is also believed to be extinct (Tye and Lau 2014).

4. From an early age, Darwin had been an avid hunter, which was put to good use during his sojourn on the HMS *Beagle*, when his hunting skills were essential for successful collecting (Dorsey 1928).

5. By this time American adventurer Victor von Hagen had already identified a farm on Santa Cruz Island as a potential location for a proposed biological station to study and protect island biota, when he recruited notable zoologist Harold Coolidge, founding director of the International Union for the Conservation of Nature (IUCN), and Julian Huxley, secretary of the Zoological Society of London, who in 1946 was to become the first director-general of the United Nations Education, Scientific, and Cultural Organization (UNESCO). After separate postwar visits to the islands, Austrian ethologist Irenäus Eibel-Eibesfeldt and American ornithologist Robert Bowman appealed to the IUCN, which received permission from Ecuador for an exploratory mission to Galápagos. After a 1957 visit to the major islands, Eibel-Eibesfeldt and Bowman submitted their reports, including plans for a research station on Santa Cruz, to the IUCN. The reports were published by UNESCO and submitted to the International Congress of Zoology in 1957, which established an international committee comprising Belgian paleontologist Victor van Straelen, Ecuadorian UNESCO delegate Luís Jaramillo, Huxley, and various influential scientists. This was to become the founding board of the Charles Darwin Foundation (CDF), a Belgian nongovernmental foundation officially recognized in 1959 by the

Ecuadorian government as administrators of Galápagos "zones of reserve and National Parks." In 1964, with funding from UNESCO, Ecuador, and various organizations, a research station was completed, scientists received medals, and the CDF was charged with official scientific advisory capacity for 25 years (Hennessy 2017).

6. A 1967 brochure by Linblad Expeditions advertised the opportunity to visit "Darwin's Galápagos" (Hennessy 2017: 87). Woram (http://www.galapagos.to/POSTAL/CACHALOT.HTM) displays a postmarked card from Santa Cruz Island of the *Golden Cachalot* on its maiden voyage in 1969 from Plymouth to the Galápagos Islands. An advertisement in the September 29, 1969 edition of *New York Magazine* (vol. 2, no. 39) by Linblad Travel offers a biweekly cruise to Darwin's fabled islands, "the land that time forgot," for a maximum of 10 passengers starting at $1,400 including airfare from New York. Another ship, the 66-passenger *Lina A*, is also mentioned as the first tour boat to visit the islands in 1969 (Salvador Ayala 2015: 58, also Brewington 2013: 107).

7. Land ownership was formalized by IERAC (Instituto Ecuatoriano de Reforma Agraria y Colonización), the Ecuadorian Institute for Agrarian Reform and Settlement, formed as part of the Land Reform, Idle Lands, and Settlement Act outlawing colonial *huasipungo* and absentee ownership, and charged with expropriating idle arable land for redistribution to farmers. IERAC was legislatively eliminated in 1994 by the Durán administration.

8. The 133 km^2 marine reserve surrounds the entire archipelago, extending approximately 65 km (40 miles) from its outermost islands. The conservation and sustainable use of marine resources are managed by the PNG and provide stipulations for human uses, including artisanal fishing, marine tourism, scientific investigation, and naval operations. In 2001, it was included in the list of World Heritage Sites, and in 2013 part of Isabela Island was added to the Annotated Ramsar List of Wetlands of International Importance (Plan de Manejo de Conservación y Uso Sustenible para la Reserva Marina de Galápagos, http://sui.ambiente.gob.ec, accessed Oct. 22, 2018).

9. The United Nations World Tourism Organization (UNWTO), which promotes responsible, sustainable, and universally accessible tourism, defines sustainable tourism as one that "takes full account of its current and future economic, social and environmental impacts, addressing the needs of visitors, the industry, the environment and host communities" (http://sdt.unwto.org/content/about-us-5, accessed October 26, 2018). Ecotourism, in particular, supports the maintenance of natural areas by generating economic benefits and providing alternative employments and income opportunities for host communities, while increasing awareness of conservation for locals and tourists alike (http://sdt.unwto.org/content/ecotourism-and-protected-areas, accessed October 26, 2018).

10. The islands are increasingly connected with the mainland and interconnected to each other. The expanding numbers of visitors, increasingly from destinations throughout the globe, and the augmented volume of air and maritime transport required to support expanding tourism are escalating the opportunities for introductions of alien exotics. External pathways for importation and internal pathways for their spread throughout the archipelago include commercial and private planes, cargo ships, tour boats, private yachts, scientific and patrol vessels, and fishing boats, both legal and illegal. The

shift toward increasing land-based tourism also augments the frequency of introductions (Toral-Granda et al. 2017). Recent estimates of waste production, mostly organic, on San Cristóbal Island range from 1,637 to 3,547 tons per year, based on 0.6 to 1.3 kg of waste per day, per person. An additional 0.8 tons of waste are generated by tour boats on a daily basis. Additional pollution is generated through the intentional and unintentional release of hydrocarbons (oil, diesel, and gas) and the accumulation of marine debris from unintentional and intentional waste disposal (Alava et al. 2014). Between 2012 and 2014, the imported cargo on ships consisted of 60% construction material, 20% dry foods and grain, 10% fresh fruits and vegetables, and 10% miscellaneous items (Toral-Granda et al. 2017: 8). Fresh organic products from mainland MegaMaxi stores are shipped in plastic (Salvador Ayala 2015: 76).

11. Salvador Ayala (2015: 79–80) documents some of the price differentials for imported fruits and vegetables, which can be more than 400% higher in the islands than in Quito. Exceptions include animal products like chicken and pork, which are in demand both locally and by tourism, and beef, which is locally produced because of the prohibition against importing fresh beef (Salvador Ayala 2015: 70).

12. Galápagos Province has the highest levels of overweight and obese adolescents (34.5%) and children (44.1%) in the country (Salvador Ayala 2015: 83).

13. Although the amount of agricultural land on the island has decreased since 2000, the number of UPAs has increased, suggesting the increased division of land parcels into smaller UPAs (Allauca et al. 2018: 21; CGREG 2016: 97).

14. Recent statistics indicate that 58% of UPAs in Galápagos were under pasturage, with 10,100 head of principally Simental cattle, the majority raised for meat production (CGREG 2016: 98).

15. In 2016, Francisco Laso undertook 39 semistructured interviews with government officials and public servants (authorities, $N = 7$), researchers and nongovernmental organizations (experts, $N = 9$), landowners, farmers, and ranchers (producers, $N = 20$), and restaurant owners (vendors, $N = 3$) on Santa Cruz and San Cristóbal Islands. Questions were motivated by increasing food insecurity, food importation, invasives, and land fragmentation and sought to gain understanding about the role of island agriculture through time, its relation with protected and invasive species, and possible future remediation. Major limitations on the establishment of local food security were identified and ranked according to their frequency of mention: lack of water; lack of labor and its relative cost; competition with imports; agricultural pests; introduced species; restrictions placed upon the importation of agricultural essentials like fertilizers, pesticides, and machinery; lack of credit or funding opportunities; lack of demand for local products; poor quality of local products; lack of technical skill; lack of transport to market; lack of collaboration between producers; conflicting interests between production and conservation; and lack of necessary infrastructure. The only issue not identified as limiting by producers was conflict with conservation goals (Laso 2016).

16. While weather conditions in Galápagos are often extreme, even on an hourly basis, decades of meteorological data from Santa Cruz Island record the increasing occurrence and severity of rainfall and thermal anomalies associated with El Niño. These

are often followed by dry La Niña conditions that can develop into severe droughts (Snell and Rea 1999).

17. A farmer interviewed by Laso (2016: 5) mentions that he was the only one not to have abandoned his land for a city job in the entire Cerro Gato region, although older residents were now returning for their retirement.

18. Between 2009 and 2011, 7,811.2–9,640.9 tons of the 16 most common fresh food products, principally transportable and durable potatoes, bananas, plantains, yuca, and onions, were annually imported to the four inhabited islands. Only 1–2% of the total tonnage was inspected. During this period, San Cristóbal was visited by 65–93 cargo ships annually (Bigue et al. 2012: table 4).

19. The most common form of infrastructure is water reservoirs, of which 885 are found on 445 UPAs in Galápagos; however, only 35 UPAs have irrigations systems, with only 17 on San Cristóbal Island (CGREG 2016: 101–102). Other organizations, including Peace Corps, the Spanish Agency of International Cooperation (AECI, Agencia Española de Cooperación Internacional), and the Israeli government, have variously provided developmental aid.

20. Vegetation change between 1987 and 2006 in San Cristóbal's ZUE is due largely to the propagation of invasive species, especially in areas of natural vegetation, artificial pastures associated with long-cycle crops, and short- and long-cycle crops. During this period, 71% of the total agricultural area (8,352.9 ha) converted to invasive species, with only 21% remaining unchanged (Villa and Segarra 2012: 86–87).

21. One example of a recent introduction to the highland areas of San Cristóbal which increasingly plagues farmers, visitors, and wildlife alike is the Carmelito, a tiny South American fly (*Philornis downsi*) that lays eggs in bird nests, where its larvae become ectoparasites of hatchlings. During the rainy season and in humid areas, large swarms of these fruit-eating flies seek blood and add to daily misery. They are commonly believed to have been named after a local lady, Doña Carmela, who accidentally introduced them in a shipment of bananas.

22. Geographically, Santa Cruz is the most central of the islands and is today the most populated (61% of island inhabitants), with the greatest range of infrastructural amenities. The current population of Galápagos includes approximately 37% who were born in the islands, while others hail principally from Guayas, Tungurahua, Manabí, Pichincha, and Loja provinces in that order, with the remainder from other provinces and a few hundred from other countries. The majority (97%) identify as mestizo, with small proportions identifying as white (2%) and of indigenous or African descent (1%). The headquarters of the PNG and CDF are located in Puerto Ayora, which has the majority of services and is the base for most tourism. The airport on Baltra Island receives more than three times more commercial passenger and cargo planes than San Cristóbal (CGREG 2016).

23. A total of 341 traditional varieties of 147 species are reported to be cultivated on 178 farms in San Cristóbal's ZUE. Many of these varieties may be local adaptations to different island ecogeographic conditions, and the source for generating new intraspecific variability. Moreover, the number of different species planted is strongly and positively

correlated with the number of traditional varieties, suggesting that agrobiodiversity is in the hands of individual farmers (Allauca et al. 2016: 39, 44).

24. The 2017 Law of Organic Agrobiodiversity, Seeds and Promotion of Sustainable Agriculture (LOASFAS, Ley Orgánica de Agrobiodiversidad, Semillas y Fomento de la Agricultura Sustentable) promotes the participation of various public, private, and academic agents to protect, revitalize, increase and invigorate agrobiodiversity within models of sustainable agriculture. Beef, chicken, eggs, milk, and cheese, along with vegetables (lettuce, carrots, peppers) grown under irrigation, are considered to have the greatest short-term success for providing island needs while diminishing reliance on mainland imports. Greenhouse tomatoes, yuca, and coffee can cover longer-term seasonal demands, and plantains, oranges, and bananas should be part of an institutional plan for longer-term improvement. Polycultural planting, especially pasturage with citrus and guava, which is practiced more pervasively on San Cristóbal than on other islands, can provide greater productivity, increased biodiversity, control of plant pathogens, soil recuperation, and carbon capture (Allauca et al. 2018; CDREG 2016). Coffee is also promising in this regard, as one farmer is experimenting with combining coffee production, shade-growing techniques, and *Scalesia* reforestation. Hacienda Tranquila on San Cristóbal is attempting to obtain empirical evidence that endemic plants can provide valuable shade for cash crops like coffee and habitat for wildlife, while simultaneously demonstrating that the protection of surrounding nesting habitat for birds and beneficial insects can provide valuable pollination and pest management for agricultural crops. Hacienda El Progreso's coffee plantation continues at La Cafetal in Socavón, whose bourbon varietals and some of the original trees planted by Cobos are marketed around the world as organic Galápagos coffee.

25. A total of 1010 artisanal fisherman are organized into four fishing cooperatives, two based on San Cristóbal (508) and one each on Santa Cruz (279) and Isabela (223) islands. Some fishermen dive exclusively for spiny lobster and sea cucumbers destined for the Asian market, some fish for demersal and pelagic fish, and others combine both activities (CGREG 2016).

26. The extent to which this is empowering fishermen, and the actual number who are participating, is perhaps debatable. However, the greatest change for island fishermen, and for agriculturalists, is through their entrance into tourism.

27. The new paintball emporium at Larry's Shooter Fire on the site of the sugar mill recently closed due to lack of interest. A zip line is located at Canopy outside of town, and across the street from the carpenter's shop is La Casa de Ceibo, boasting the world's largest ceiba tree, which is open for visits and parties. The Biological Station-Private Reserve operated by Jatun Sacha, a Quito-based NGO, accepts volunteers who contribute to agricultural and touristic efforts. Hacienda La Tranquila at La Soledad also hosts volunteers engaged in a variety of ecological initiatives, especially the eradication of invasives and replanting of endemics. The extent to which El Progreso derives significant benefits from these venues is debatable.

28. The house site is located on private property but is registered as a historic heritage site (IBI-20-01-51-000-000002) with the National Cultural Heritage Institute of Ecuador (INPC, Instituto Nacional de Patrimonio Cultural–Ecuador). In August 2018, the junta

parroquial of the autonomous government of El Progreso petitioned the INPC to register the entire *ingenio* property below the house site, its preserved infrastructure, and the Cobos tomb site, along with house site as a historic industrial complex of elevated importance to national Ecuadorian cultural heritage. As of May 2019, a technical visit is being planned to the site by various officials, including Florencio Delgado, to delimit its boundaries and implement the petition for conservation.

REFERENCES

Abbott, Elizabeth

2008 *Sugar: A Bittersweet History*. Duckworth Overlook, London.

Adams, William H.

2002 Machine Cut Nails and Wire Nails: American Production and Use for Dating 19th-Century and Early-20th-Century Sites. *Historical Archaeology* 36(4): 66–88.

Agassiz, Alexander

1892 General Sketch of the Expedition of the "Albatross," from February to May, 1891. *Bulletin of the Museum of Comparative Zoölogy* 23: 1–89.

Agassiz, G. R. (editor)

1913 *Letters and Recollections of Alexander Agassiz with a Sketch of his Life and Work*. Houghton Mifflin, Boston.

Ahumada Escobar, Catalina

2010 Los Apuntes y Cuentas de la Hacienda Coconuco (1770–1850). Las Prácticas de Leer y Escribir en una Hacienda Esclavista Colombiana. *Historia Crítica* 40(42): 36–53.

Alava J. J., C. Palomera, L. Bendell, and P. S. Ross

2014 Pollution as an Emerging Threat for the Conservation of the Galapagos Marine Reserve: Environmental Impacts and Management Perspectives. In *The Galapagos Marine Reserve. Social and Ecological Interactions in the Galapagos Islands*, edited by J. Denkinger and L. Vinueza, pp. 247–282. Springer, Cham.

Alexandre, A., J.-D. Meunier, A.-M Lezine, A. Vincens, and D. Schwartz

1997 Phytoliths: Indicators of Grassland Dynamics during the Late Holocene in Intertropical Africa. *Palaeogeography, Palaeoclimatology, Palaeoecology* 136(1–4): 213–229.

Alexandre, Anne, Jean-Dominique Meunier, Andre Mariotti, and Francois Soubies

1999 Late Holocene Phytolith and Carbon-Isotope Record from a Latosol at Salitre, South-Central Brazil. *Quaternary Research* 51: 187–194.

Allauca, J., M. Valverde, and C. Tapia

2018 Conocimiento, Manejo y Uso de la Agroboversidad en la Isla San Cristóbal. *INIAP. Boletín Técnico* 173. INDIGO 480, Puerto Baquerizo Moreno.

Amaral, Samuel
2002 *The Rise of Capitalism on the Pampas: The Estancias of Buenos Aires, 1785–1870.* Cambridge University Press, Cambridge.
Ameringer, Charles D.
1963 The Panama Canal Lobby of Philippe Bunau-Varilla and William Nelson Cromwell. *American Historical Review* 68: 346–363.
Anderson, Atholl J.
2002 Faunal collapse, landscape change and settlement history in remote Oceania. *World Archaeology* 33: 375–390.
Anderson, Atholl, Karen Stothert, Helene Martinsson-Wallin, Paul Wallin, Iona Flett, Simon Haberle, Henk Heijnis, and Edward Rhodes
2016 Reconsidering Precolumbian Human Colonization in the Galápagos Islands, Republic of Ecuador. *Latin American Antiquity* 27: 169–183.
Andrade, Margot
2012 *La Colombie et la France: Relations culturelles XIXe–XXe siècles.* L'Harmattan, Paris.
Angermeyer, Johanna
1989 *My Father's Island: A Galapagos Quest.* Viking, New York.
Anonymous
1866 *The Manufacturing Interest of the City of Buffalo, Including Sketches of the History of Buffalo with Notice of its Principal Manufacturing Establishments* 2nd Edition. C.F.S. Thomas, Buffalo.
1893 Improved Triple Effect Evaporator. *Scientific American* 68(15): 228.
1897 The Development of Sugar Machinery in Glasgow. *The Sugar Cane* 29: 133–136.
Araújo, Astolfo Gomes de Mello, and Marcos Rogério Ribeiro de Carvalho
1993 A Louça Inglesa do Século XIX: Considerações sobre a Terminologia e Metodologia Utilizadas no Sítio Flôrencio de Abreu. *Revista do Museu de Arqueologia e Etnologia* 3: 81–95.
Arthur, Karen E., Michelle C. Boyle, and Colin J. Limpus
2008 Ontogenetic Changes in Diet and Habitat Use in Green Sea Turtle (*Chelonia mydas*) Life History. *Marine Ecology Progress Series* 362: 303–311.
Astudillo, Fernando J.
2017 *Environmental Historical Archaeology of the Galápagos Islands: Paleoethnobotany of Hacienda El Progreso, 1870–1904.* Unpublished Ph.D. dissertation, Department of Archaeology, Simon Fraser University, Burnaby.
2018a Soil Phytoliths as Indicators of Initial Human Impact on San Cristóbal Island, Galápagos. *Palaeogeography, Palaeoclimatology, Palaeoecology* 490: 522–532.
2018b Illustrated Catalogue of Phytoliths from Modern Plants of the Galápagos Islands: Economic Species of San Cristóbal Island. *Avances,* in press.
2018c Environmental and Historical Archaeology of the Galápagos Islands: Archaeobotany of El Progreso, 1870–1920. *Vegetation History and Archaeobotany* https://doi.org/10.1007/s00334-018-0668-9

Athens, Stephen J., H. David Tuggle, Jerome V. Ward, and David J. Welch
2002 Avifaunal Extinctions, Vegetation Change, and Polynesian Impacts in Prehistoric Hawai'i. *Archaeology of Oceania* 37: 57–78.
Atkinson, Rachel, Mark Gardener, Grant Harper, and Victor Carrion
2012 Fifty Years of Eradication as a Conservation Tool in Galápagos. What are the Limits? In *The Role of Science for Conservation*, edited by M. Wolff and M. Gardener, pp. 183–198. Routledge, London.
Ayres, Philip
1684 *The Voyages and Adventures of Capt. Barth. Sharp and others, in the South Sea: being a Journal of the Same*. P.A. Esq., London.
Baker, Charles H., Jr.
1951 *The South American Gentleman's Companion. Vol. 2, Being an Exotic Drinking Book or, Up and Down the Andes with Jigger, Beaker & Flask*. Crown Publishers, New York.
Balée, William
1998 Historical Ecology: Premises and Postulates. In *Advances in Historical Ecology*, edited by William Balée, pp. 13–29. Columbia University Press, New York.
2006 The research program of historical ecology. *Annual Review of Anthropology* 35: 75–98.
Balée, William, and Clark L. Erickson
2006 Time, Complexity, and Historical Ecology. In *Time and Complexity in Historical Ecology*, edited by W. Balée and C. L. Erickson, pp. 1–17. Columbia University Press, New York.
Barboni, Doris, Laurent Bremond, and Raymonde Bonnefille
2007 Comparative study of modern phytolith assemblages from inter-tropical Africa. *Palaeogeography, Palaeoclimatology, Palaeoecology* 246: 454–470
Barnett, Bruce D., and Robert L. Rudd
1983 Feral Dogs of the Galápagos Islands: Impact and Control. *Journal for the Study of Animal Problems* 4: 44–58.
Barrett, Ward
1965 Caribbean Sugar Production Standards in the Seventeenth and Eighteenth Centuries. In *Merchants and Scholars*, edited by John Parker, pp. 145–170. University of Minnesota Press, Minneapolis.
1970 *The Sugar Hacienda of the Marques del Valle*. University of Minnesota Press, Minneapolis.
Barton, Amanda Marie
2012 An Archaeological and Historical Investigation of a 19th Century Leprosarium at Hassel Island, St. Thomas, U.S. Virgin Islands. Unpublished MA thesis, Dept. of Anthropology, University of Tennessee, Knoxville.
Bassett, Carol Ann
2009 *Galápagos at the Crossroads: Pirates, Biologists, Tourists, and Creationists Battle for Darwin's Cradle of Evolution*. National Geographic, Washington, D.C.

Bauer, Arnold J.
2001 *Goods, Power, History: Latin America's Material Culture*. Cambridge University Press, Cambridge.
Baur, Georg
1891 The Galápagos Islands. *Proceedings of the American Antiquarian Society* 7: 418–423.
Baxter, Jane Eva
2005 *The Archaeology of Childhood: Children, Gender, and Material Culture*. AltaMira Press, Walnut Creek, CA.
Beaudry, Mary Carolyn
2006 *Findings: The Material Culture of Needlework and Sewing*. Yale University Press, New Haven.
Becerra Chaves, Nicole A.
2017 Development of a Water Filtration System Using Design Thinking for Social Innovation in San Cristobal, Galapagos Ecuador. College of Science and Engineering, Universidad San Francisco de Quito, Quito.
Becher, Pablo Ariel
2011 Patrimonio urbano y arqueología histórica en el Barrio Noroeste de la ciudad de Bahía Blanca (fines de siglo XIX–principios del XX). Universidad Nacional del Sur, Bahia Blanca, Argentina.
Beck, Rollo
1906 Rollo Beck Galapagos Expedition Journal. California Academy of Sciences. http://www.biodiversitylibrary.org/bibliography/59676
Beckman, Ericka
2013 *Capital Fictions: The Literature of Latin America's Export Age*. University of Minnesota Press, Minneapolis.
Beebe, William
1924 *Galapagos: World's End*. G.P. Putnam's Son, New York.
Berlanga, Tomás de
1884[1535] Carta de Su Magestad de Fray Tomás de Berlanga, Describiendo su Viaje desde Panamá a Puerto Viejo, e los Trabajos que Padeció en la Navegación. In *Colección de Documentos Inéditos Relativos de las Antiguas Posesiones Españoles de América y Oceanía Sacados de los Archivos del Reino y Muy Especialmente del de Indias,* Vol. 41, edited by Joaquin F. Pacheco, Francisco de Cárdenas, and Luís Torres de Mendoza, pp. 538–544. Manuel G. Hernández, Madrid.
Berthollet, C. L., and A. B. Berthollet
1841 *Elements of the Art of Dyeing and Bleaching*. Thomas Tegg, London.
Bigue, Marcel, Laura Brewington, Kleber Cervantes, and Oswaldo Rosero
2012 *The Quarantine Chain: Establishing an Effective Biosecurity System to Prevent the Introduction of Invasive Species into the Galápago Islands*. WildAid, San Francisco.
Bilbao, Manuel
1865 *El Pirata del Guayas*. Tornero e Hijos, Valparaiso.

Blake, S., A. Guézou, S. L. Deem, C. B. Yackulic, and F. Cabrera

2015 The Dominance of Introduced Plant Species in the Diets of Migratory Galapagos Tortoises Increases with Elevation on a Human-occupied Island. *Biotropica* 47: 246–258.

Bognoly, José A., and José Moisés Espinosa

1905 *Las Islas Encantadas ó el Archipiélago de Colón*. Imprenta y Litografía del Comercio, Guayaquil.

Bolaños, Mónica, María Moreira, Rocío Murillo, Guillermo Sevilla, Marco Vargas, Marco Rosero, and Leonardo Arcos

2002 *Informe Técnico Final: Investigación y Puesta en Valor Casa de Hacienda y Parte del Ingenio "El Progreso."* Instituto Nacional de Patrimonio Cultural del Ecuador, Quito.

Bolten, Alan B.

2003 Variation in Sea Turtle Life History Patterns: Neritic vs. Oceanic Developmental Stages. In *The Biology of Sea Turtles* Vol. 2, pp. 243–257, edited by P. L. Lutz, J. A. Musick and J. Wyneken. CRC, Boca Raton.

Borja, Rafael

1948 *Galápagos: Mito y Realidad*. Casa de la Cultura, Quito.

Bradley, Peter T.

1989 *The Lure of Peru: Maritime Intrusion into the South Sea, 1598–1701*. MacMillan, Houndsmills.

2015 *The Last Buccaneers in the South Sea 1688–1695*. Lulu.com, Milton Keynes.

Braje, Todd J., and Jon M. Erlandson

2013 Human Acceleration of Animal and Plant Extinctions: A Late Pleistocene, Holocene, and Anthropocene Continuum. *Anthropocene* 4: 14–23.

Brantlinger, Patrick

2018 *Barbed Wire: Capitalism and the Enclosure of the Commons*. Routledge, Abingdon.

Bremond, Laurent, Anne Alexandre, Christelle Hely, and Joel Guiot

2005 A Phytolith Index as a Proxy of Tree Cover Density in Tropical Areas: Calibration with Leaf Area Index along a Forest-Savanna Transect in Southeastern Cameroon. *Global and Planetary Change* 45(4): 277–293.

Brewington, Laura

2011 Agriculture and Conservation in the Galápagos Islands. *Grassroots Development* 32(1): 54–59.

2013 The Double Bind of Tourism in Galapagos Society. In *Science and Conservation in the Galapagos Islands*, edited by S. J. Walsh and C. F. Mena, pp. 105–125. Springer Science+Business Media, New York.

Bridgewater, Peter, and Eric S. Higgs

2011 Engaging with Novel Ecosystems. *Frontiers in Ecology and the Environment* 9: 423.

Briggs, John C.

2011 Marine Extinctions and Fish Conservation. *Marine Biology* 158: 485–488.

Brink, J. W.
1997 Fat Content in Leg Bones of *Bison bison*, and Applications to Archaeology. *Journal of Archaeological Science* 24: 259–274.
Bromley, Ray
1981 The Colonization of Humid Tropical Areas in Ecuador. *Singapore Journal of Tropical Geography* 2: 15–26.
Brooks, Alasdair, and Ana Cristina Rodríguez Y.
2012 A Venezuelan Household Clearance Assemblage of 19th-Century British Ceramics in International Perspective. *Post-Medieval Archaeology* 46(1): 70–88.
Bull, Gail, and Sebastien Payne
1982 Tooth Eruption and Epiphyseal Fusion in Pigs and Wild Boar. In *Ageing and Sexing Animal Bones from Archaeological Sites*, edited by B. Wilson, C. Grigson, and S. Payne, pp. 55–71. British Series 109, B.A.R., Oxford.
Bullock, D., and J. Rackham
1982 Epiphyseal Fusion and Tooth Eruption of Feral Goats from Moffatdale, Dumfries and Galloway, Scotland. In *Ageing and Sexing Animal Bones from Archaeological Sites*, edited by B. Wilson, C. Grigson, and S. Payne, pp. 73–80. British Series 109, B.A.R., Oxford.
Bunker, Steven B.
2012 *Creating Mexican Consumer Culture in the Age of Porfirio Díaz.* University of New Mexico Press, Albuquerque.
Burgos-Paz, W., C. A. Souza, H. J. Megans, et al.
2013 Porcine Colonization of the Americas: A 60K SNP Story. *Heredity* 110: 321–330.
Burney, D., and L. P. Burney
2007 Paleoecology and "inter-situ" restoration on Kaua'i, Hawai'i. *Frontiers in Ecology and the Environment* 5: 483–490.
Burney, D., and Timothy Flannery
2005 Fifty millennia of catastrophic extinctions after human contact. *Trends in Ecology and Evolution* 20: 395–401.
Burney, James
1813–1817 *A Chronological History of the Discoveries in the South Sea or Pacific Ocean.* Vol. 4. L. Hansard, London.
Bush, Mark B., Alejandra Restrepo, and Aaron F. Collins
2014 Galápagos History, Restoration, and a Shifted Baseline. *Restoration Ecology* 22: 296–298.
Cai, W., S. Borlace, M. Lengaigne, P. van Rensch, M. Collins, A. Timmerman, A. Santoso, M. J. McPhaden, L. Wu, M. H. England, E. Guilyardi, and F. Jin
2014 Increasing frequency of extreme El Niño events due to greenhouse warming. *Nature Climate Change* 4: 111–116.
California Academy of Sciences
1907 *Expedition of the California Academy of Sciences to the Galapagos Islands, 1905–1906.* California Academy of Sciences, San Francisco.

Calvo Rebollar, Miguel
2009 Dinero no veían, solo fichas. El pago de salarios en las salitreras de Chile hasta 1925. *De Re Metallica* 12: 9–30.
Campbell, Karl, C. Josh Donlan, Felipe Cruz, and Victor Carrion
2004 Eradication of Feral Goats *Capra hircus* from Pinta Island, Galápagos, Ecuador. *Oryx* 38: 328–333.
Capello, Ernesto
2011 *City at the Center of the World: Space, History, and Modernity in Quito*. University of Pittsburgh Press, Pittsburgh.
Carey, David, Jr.
2014 Drunks and Dictators: Inebriation's Gendered, Ethnic, and Class Components in Guatemala, 1898–1944. In *Alcohol in Latin America: A social and cultural history*, edited by Gretchen Pierce and Áurea Toxqui, pp. 131–157. University of Arizona Press, Tucson.
Carmi, Ore, C. C. Witt, A. Jaramillo, and J. P. Dumbacher
2016 Phylogeography of the Vermilion Flycatcher species complex: Multiple speciation events, shifts in migratory behavior, and an apparent extinction of a Galápagos-endemic bird species. *Molecular Phylogenetics and Evolution* 102: 152–173. http://dx.doi.org/10.1016/j.ympev.2016.05.029
Carrion, Victor, C. Josh Donlan, Karl Campbell, Christian Lavoie, and Felipe Cruz
2007 Feral Donkey (*Equus asinus*) Eradications in the Galápagos. *Biodiversity and Conservation* 16: 437–445.
2011 Archipelago-wide Island Restoration in the Galápagos Islands: Reducing Costs of Invasive Mammal Eradication Programs and Reinvasion Risk. *PLOS ONE* 6(5): 1–7.
Causton, C. E., S. B. Peck, B. J. Sinclair, L. Roque-Albelo, C. J. Hodgson, and B. Landry
2006 Alien Insects: Threats and Implications for Conservation of Galápagos Islands. *Annals of the Entomological Society of America* 99: 121–143.
Cevallos-Falquez, Orly, Cecilio Barba, Juan Vicent Delgado, Ana González, José Perea, Elena Angón, and Antón García
2016 Caracterización Zoométrica y Morfológia del Ganado Criollo de Manabí (Ecuador). *Revista Científica (Universidad de Zulia Facultad de Ciencias Veterinarias División de Investigación)* 26: 313–323.
CGREG (Consejo de Gobierno del Régimen Especial de Galápagos)
2011 *Plan Estratégico de la Junta Parroquial El Progreso*. http://www.gobiernogalapagos.gob.ec/wp-content/uploads/downloads/2013/08/Plan-Desarrollo-Progreso.pdf
2016 Plan de Desarrollo Sustentable y Ordenamiento Territorial del Régimen Especial de Galápagos. Plan Galápagos. Puerto Baquerizo Moreno, Galápagos, Ecuador.
Chambers, Steven M., and David W. Steadman
1986 Holocene Terrestrial Gastropod Faunas from Isla Santa Cruz and Isla Floreana: Evidence for Late Holocene Declines. *Transactions of the San Diego Society of Natural History* 21(6): 89–110.

Chaves, Jaime A., Micaela Peña, Jhonnattan A. Valdés-Uribe, Juan P. Muñoz-Pérez, Felipe Vallejo, Maike Heidemeyer, and Omar Torres-Carvajal

2017 Connectivity, Population Structure, and Conservation of Ecuadorian Green Sea Turtles. *Endangered Species Research* 32: 251–264.

Chiriboga Vega, Manuel

2013 *Jornaleros, grandes propietarios y exportación cacaotera, 1790–1925*. Corporación Editora Nacional, Quito.

Church, George Earl

1983 *Ecuador in 1881*. Government Printing Office, Washington, D.C.

Clayton, Jane M.

2014 *An Alphabetical List of Ships Employed in the South Sea Whale Fishery from Britain: 1775–1815*. Berforts Information Press, Hastings.

Clissold, Stephen

1954 *Conquistador: The Life of Don Pedro Sarmiento de Gamboa*. Derek Verschoyle, London.

Coblentz, Bruce E.

1978 The Effects of Feral Goats (*Capra hircus*) on Island Ecosystems. *Biological Conservation* 13: 279–285.

Coblentz, Bruce E., and D. W. Baber

1987 Biology and Control of Feral Pigs on Isla Santiago, Galápagos, Ecuador. *Journal of Applied Ecology* 24: 403–418.

Coe, Heloisa H. G., Anne Alexandre, Cacilda N. Carvalho, Guaciara M. Santos, Antonio S. da Silva, Leandro O. F. Sousa, and Igo F. Lepsch

2013 Changes in Holocene tree cover density in Cabo Frio (Rio de Janeiro, Brazil): Evidence from soil phytolith assemblages. *Quaternary International* 287: 63–72.

Coe, Heloisa H. G., Kita Macario, Jenifer G. Gomes, Karina F. Chueng, Fabiana Oliveira, Paulo R. S. Gomes, Carla Carvalho, Roberto Linares, Eduardo Alves, and Guaciara M. Santos

2014 Understanding Holocene variations in the vegetation of Sao Joao River basin, southeastern coast of Brazil, using phytolith and carbon isotopic analyses. *Palaeogeography, Palaeoclimatology, Palaeoecology* 415: 59–68.

Coe, Heloisa Helena Gomes, Yame Bronze Medina Ramos, Catia Pereira dos Santos, Andre Luiz Carvalho da Silva, Carolina Pereira Silvestre, Natalia Borrelli, and Leandro de Oliveira Furtado de Sousa

2015 Dynamics of production and accumulation of phytolith assemblages in the Restinga of Marica, Rio De Janeiro, Brazil. *Quaternary International* 434(B): 58–69.

Colinvaux, Paul A., and Eileen K. Schofield

1996 Historical Ecology in the Galapagos Islands: 1. A Holocene Pollen Record from El Junco Lake, Isla San Cristobal. *Journal of Ecology* 64: 989–1012.

Collier, Marcus J., and Catherin Devitt

2016 Novel Ecosystems: Challenges and Opportunities for the Anthropocene. *Anthropocene Review* 3: 231–242.

Collins, A., and M. B. Bush
2011 An analysis of modern pollen representation and climatic conditions on the Galápagos Islands. *Holocene* 21(2): 237–250
Colnett, James
1798 *A Voyage to the South Atlantic and Round Cape Horn into the Pacific Ocean, for the Purpose of Extending the Spermaceti Whale Fisheries, and other Objects of Commerce, by Ascertaining the Ports, Bays, Harbours, and Anchoring Births, in Certain Islands and Coasts in the Seas at which the Ships of the British Merchants Might be Refitted.* W. Bennett, Westminster.
Compañía Guía del Ecuador
1909 *El Ecuador: Guia Comercial, Agricola, e Industrial de la Republica.* E. Rodenas, Guayaquil.
Conrad, Cyler, Laura Pagès Barceló, Jeffrey A. Seminoff, Calandra Turner Tomaszewicz, Marie Labonte, Brian M. Kemp, Emily Lena Jones, Michael Stoyka, Kale Bruner, and Allen Pastron
2018 Ancient DNA Analysis and Stable Isotope Ecology of Sea Turtle (Cheloniidae) from the Gold Rush Era (1850s) Eastern Pacific Ocean. *Open Quaternary* 4(3): 1–13. https://doi.org/10.5334/oq.41
Conrad, Cyler, and Allen Pastron
2014 Galapagos Tortoises and Sea Turtles in Gold Rush-Era California. *California History* 91: 20–39.
Conway, Ainslee, and Frances Conway
1947 *The Enchanted Islands.* G.P. Putnam's Sons, New York.
Coulter, John
1845 *Adventures in the Pacific; with Observations on the Natural Productions, Manners and Customs of the Natives of the Various Islands; Together with Remarks on Missionaries, British and Other Residents, etc. etc.* William Curry, Jun. and Company, Dublin.
Cowley, William Ambrosia
1699 *Capt. Cowley's Voyage Round the World.* In *A Collection of Original Voyages*, edited by W. Hacke, Chapters 3–4. James Knapton, London.
Crealock, William I. B.
1955 *Cloud of Islands: By Sail to the South Seas.* Hastings House, New York.
Crosby, Alfred W.
1986 *Ecological Imperialism: The Biological Expansion of Europe, 900–1900.* Cambridge University, Cambridge.
Cruz, Felipe, C. Josh Donlan, Karl Campbell, and Victor Carrion
2005 Conservation Action in the Galápagos: Feral Pig (*Sus scrofa*) eradication from Santiago Island. *Biological Conservation* 121: 473–478.
Cuevas, Jaime G., and Carlos Le Quesne
2006 Low Vegetation Recovery after Short-term Cattle Exclusion on Robinson Crusoe Island. *Plant Ecology* 183: 105–124.

Curchamp, F., Jean-Louis Chapuis, and Michael Pascal
2003 Mammal invaders on islands: impact, control and control impact. *Biological Review* 78: 347–383.
Daily Evening Bulletin
1871a The Lower California Imbroglio. *Daily Evening Bulletin* (San Francisco). Issue 59, Dec. 16, 1871: n.p. *19th Century U.S. Newspapers*. Web February 15, 2017.
1871b The Magdalena Bay Imbroglio. *Daily Evening Bulletin* (San Francisco). Issue 62, Dec. 20, 1871: n.p. *19th Century U.S. Newspapers*. Web February 15, 2017.
1871c The Magdalena Bay Imbroglio. *Daily Evening Bulletin* (San Francisco, California). Issue 63, Dec. 21, 1871: n.p. *19th Century U.S. Newspapers*. Web February 15, 2017.
1872a The Lower California Imbroglio. *Daily Evening Bulletin* (San Francisco). Issue 19, April 29, 1872: n.p. *19th Century U.S. Newspapers*. Web February 15, 2017.
1872b The Magdalena Bay Imbroglio. *Daily Evening Bulletin* (San Francisco). Issue 59, Jan. 3, 1872: n.p. *19th Century U.S. Newspapers*. Web February 15, 2017.
1874 Lower California Again in Court. *Daily Evening Bulletin* (San Francisco). Issue 62, June 18, 1874: n.p. *19th Century U.S. Newspapers*. Web February 15, 2017.
Dale, Ronald
1977 *The Price Guide to Black and White Pot-Lids*. Antique Collectors Club, Woodbridge, UK.
Daly, Kath
1989 Eradication of Feral Goats from Small Islands. *Oryx* 23: 71–75.
Dampier, William
1697 *A New Voyage Around the World*. James Knapton, London.
Darwin, Charles
1845 *Journal of Researches into the Natural History and Geology of the Countries Visited During the Voyage of H.M.S. Beagle Round the World under the Command of Capt. Fitz Roy, R.N.* Second edition, with additions. John Murray, London.
Decauville, M.
1884 Portable Railways. *Scientific American Supplement* 446, July 19.
Deerr, Noël
1911 *Cane Sugar: A Textbook on the Agriculture of the Sugar Cane, the Manufacture of Cane Sugar, and the Analysis of Sugar House Products Together with a Chapter on the Fermentation of Molasses*. Norman Rodger, Altrincham.
Deerr, Noël, and Alexander Brooks
1940 The Early Use of Steam Power in the Sugar Cane Industry. *Transactions of the Newcomen Society* 21(1): 11–21.
De Kay, Drake
1871 *In Relation to Occurrences at Magdalena Bay, Lower California, Mexico. 1871*. E. Bosquil, San Francisco.
Delano, Amasa
1817 *Narrative of Voyages and Travels in the Northern and Southern Hemispheres Comprising Three Voyages Round the World*. E. G. House, Boston.

Derevianko, Anatoly P., Alexander V. Postnov, Eugeny P. Rybin, Yaroslav V. Kuzmin, and Susan G. Keates

2005 The Pleistocene Peopling Of Siberia: A Review of Environmental and Behavioural Aspects. *Indo-Pacific Prehistory Association Bulletin* 25: 57–68.

Desender, Konjev, Leon Baert, Jean-Pierre Maelfait, and Peter Verdyck

1999 Conservation on Volcán Alcedo (Galápagos): Terrestrial Invertebrates and the Impact of Feral Goats. *Biological Conservation* 87: 303–310.

Doak, Daniel F., Victoria J. Bakker, Bruce Evan Goldstein, and Benjamin Hale

2014 What is the Future of Conservation? *Trends in Ecology and Evolution* 29: 77–81.

Dolin, Eric Jay

2007 *Leviathan: The History of Whaling in America.* W.W. Norton, New York.

Donoso Bustamente

2012 *Piratas en Galápagos (1680–1720).* Editorial Ecuador, Quito.

Dorsey, George A.

1928 *The Evolution of Charles Darwin.* George Allen and Unwin, London.

D'Orso, Michael

2002 *Plundering Paradise. The Hand of Man on the Galápagos Islands.* Harper Collins, New York.

Dowler, Robert C., Darin S. Carroll, and Cody W. Edwards

2000 Rediscovery of Rodents (Genus *Nesoryzomys*) Considered Extinct in the Galápagos Islands. *Oryx* 34: 109–117.

Ducéré, Edouard

1895 Journal de Bord D'un flibustier (1686–1693). D'aprés un Manuscrit de la Bibliothèque Nationale. *Bulletin de la Société des Sciences et Arts de Bayonne.* Lamaignère, Bayonne.

Duffy, David C.

1981 Ferals that Failed. *Noticias de Galápagos* 33: 21–22.

Durham, William H.

2012 What Darwin Found Convincing in Galápagos. In *The Role of Science for Conservation*, edited by M. Wolff and M. Gardener, pp. 3–15. Routledge, London.

Eckhardt, Robert C.

1972 Introduced Plants and Animals in the Galápagos Islands. *BioScience* 22: 385–390.

Egnal, George

2015 *The Bloody Human Saga of Galápagos.* BoD, Norderstedt.

Elkin, Dolores, Amaru Argüeso, Mónica Grosso, Cristian Murray, Damián Vainstub, Ricardo Bastida, and Virginia Dellino-Musgrave

2006 Archaeological research on HMS *Swift*: a British Sloop of War lost off Patagonia, Southern Argentina, in 1770. *International Journal of Nautical Archaeology* 36(1): 32–58.

Epler, Bruce

2007 *Tourism, the Economy, Population Growth, and Conservation in Galapagos.* Charles Darwin Foundation, Puerto Ayora, Ecuador.

2013 *Galápagos: A Human History.* Fast Pencil, Campbell, CA.

Estrada, Jenny

1988 *La Balsa en la Historia de la Navegación Ecuatoriana. Compilación de Crónicas, Gráficas y Testimonios.* Instituto de Historia Marítima, Guayaquil.

Evett, Rand R., Arthur Dawson, and James W. Bartolome

2012 Estimating Vegetation Reference Conditions by Combining Historical Source Analysis and Soil Phytolith Analysis at Pepperwood Preserve, Northern California Coast Ranges, U.S.A. *Restoration Ecology* 21(4): 1–10

Evett, Rand R., Ernesto Franco-Vizcaino, and Scott L. Stephens

2007 Phytolith Evidence for the Absence of a Prehistoric Grass Understory in a Jeffrey Pine-Mixed Conifer Forest in the Sierra San Pedro Martir, Mexico. *Canadian Journal of Forest Research* 37(2): 306–317.

Eyer, Marguerite

1918 French Expansion into the Pacific in the 17th, 18th, and 19th Centuries. *Annual Publication of the Historical Society of Southern California* 11(1): 5–23.

Federal Writers Project of the Works Progress Administration of Massachusetts

1987 *Whaling Masters.* Stokvis Studies in Historical Chronology and Thought, Number 8. Borgo Press, San Bernardino, CA.

Feister, Lois M.

2009 The Orphanage at Schuyler Mansion. In *The Archaeology of Institutional Life*, edited by April M. Beisaw and James G. Gibb, pp. 105–116. University of Alabama Press, Tuscaloosa.

Fike, Richard E.

1987 *The Bottle Book: A Comprehensive Guide to Historic, Embossed Medicine Bottles.* Peregrine Smith, Salt Lake City.

Fishkis, O., J. Ingwersen, and T. Streck

2009 Phytolith Transport in Sandy Sediment: Experiments and Modeling. *Geoderma* 151: 168–178.

Flett, Iona, and Simon Haberle

2008 Traces of Human Impact in the Far Eastern Pacific. In *Islands of Inquiry: Colonisation, Seafaring and the Archaeology of Maritime Landscapes*, edited by G. Clark, F. Leach, and S. O'Conner, 281–300. Australian National University Press, Canberra.

Foutiou, Evgenia

2012 Working with "La Medicina": Elements of Healing in Contemporary Ayahuasca Rituals. *Anthropology of Consciousness* 23(1): 6–27.

Fowler de Neira, Lynn E., and Mark K. Johnson

1985 Diets of Tortoises and Feral Burros on Volcan Alcedo, Galápagos. *Journal of Wildlife Management* 44: 165–169.

Fraser, Janice

2002 *The Domestic Front: An Archaeological Investigation of the Albert Barracks Ceramics Assemblage.* BA Honours Thesis, Department of Anthropology, University of Auckland.

Froese, R., and D. Pawley (editors)

2017 FishBase. World Wide Web electronic publishers. www.fishbase.org (10/2017).

Gaitán Ammann, Felipe

2005 With a hint of Paris in the mouth: fetishized toothbrushes or the sensuous experience of modernity in late 19th century Bogotá. In *Archaeologies of Materiality*, edited by Lynn Meskell, pp. 71–95. Blackwell, Oxford.

2011 A Life on Broken China: Figuring Senses of Capitalism in Late Nineteenth-Century Bogotá. In *The Archaeology of Capitalism in Colonial Contexts*, edited by Sarah K. Croucher and Lindsay Weiss, pp. 143–163. Springer, New York.

Galloway, J. H.

1989 *The Sugar Cane Industry: An Historical Geography from its Origins to 1914*. Cambridge University Press, Cambridge.

Gardener, Mark R., Rachel Atkinson, and Jorge Luis Rentería

2010 Eradications and People: Lessons from the Plant Eradication Program in Galapagos. *Restoration Ecology* 18: 20–29.

Gebhardt, G. F.

1928 *Steam Power Plant Engineering*. Sixth Edition. John Wiley and Sons, New York.

Geerlings, H. C. Prinsen

1912 *The World's Sugar Industry: Past and Present*. Cambridge University Press, New York.

Geist, Dennis J., Alexander McBirney, and Robert A. Duncan

1986 Geology and Petrogenesis of Lavas from San Cristobal Island, Galapagos Archipelago. *Geological Society of America Bulletin* 97: 555–566.

Gerhard, Peter

1960 *Pirates on the West Coast of New Spain 1575–1742*. Arthur H. Clark, Glendale, CA.

Gibbons, Ann

2000 The Peopling of the Pacific. *Science* 291: 1735–1737.

Gilliss, James Melville

1855 *Chile: Its Geography, Climate, Earthquakes, Government, Social Conditions, Mineral and Agricultural Resources, Commerce, &c. &c.* A.O.P. Nicholson, Washington.

Ginja, C., L. T. Gama, A. Martínez, N. Sevane, et al.

2017 Genetic Diversity and Patterns of Population Structure in Creole Goats from the Americas. *Animal Genetics* 48: 315–329.

Giovannetti, Marco Antonio, and Verónica Lema

2007 Circulación de Bienes y Patrones de Descarte En La Estancia Iraola. *Comechingonia Virtual: Revista Electrónica de Arqueología* 1(2): 84–105.

Godden, Geoffrey A.

1964 *Encyclopaedia of British Pottery and Porcelain Marks*. London, Barrie and Jenkins.

Gonzales, Michael J.

1985 *Plantation Agriculture and Social Control in Northern Peru, 1875–1933*. University of Texas, Austin.

González, J. A., C. Montes, J. Rodríguez, and W. Tapia

2008 Rethinking the Galapagos Islands as a complex social-ecological system: implications for conservation and management. *Ecology and Society* 13(2): 13. http://www.ecologyandsociety.org/vol13/iss2/art13/

Grant, Annie

1982 The Use of Tooth Wear as a Guide to the Age of Domestic Ungulates. In *Ageing and Sexing Animal Bones from Archaeological Sites*, edited by B. Wilson, C. Grigson, and S. Payne, pp. 91–108. British Series 109, B.A.R., Oxford.

Grant, Jonathan A.

2007 *Rulers, Guns, and Money: The Global Arms Trade in the Age of Imperialism.* Harvard University Press, Cambridge.

Grant, K. Thalia, and Gregory B. Estes

2009 *Darwin in Galápagos: Footsteps to a New World.* Princeton University Press, Princeton.

Grayson, Donald K.

2001 The Archaeological Record of Human Impacts on Animal Populations. *Journal of World Prehistory* 15: 1–68.

Grehan, John

2001 Biogeography and Evolution of the Galapagos: Integration of the Biological and Geological Evidence. *Biological Journal of the Linnean Society* 74: 267–287.

Grigson, Caroline

1982 Sex and Age Determination of Some Bones and Teeth of Domestic Cattle: A Review of the Literature. In *Ageing and Sexing Animal Bones from Archaeological Sites*, edited by B. Wilson, C. Grigson, and S. Payne, pp. 7–23. British Series 109, B.A.R., Oxford.

Guerrero, Andrés

1997 The Construction of a Ventriloquist's Image: Liberal Discourse and the "Miserable Indian Race" in Late-Nineteenth-Century Ecuador. *Journal of Latin American Studies* 29(3): 555–590.

Guevara Ruíz, Carlos Enrique

2015 *Rebelión en Galápagos: La Hacienda "Progreso" de Manuel J. Cobos en la Isla San Cristóbal, 1879–1904.* Universidad Andina Simón Bolívar, Sede Ecuador, Quito.

Guézou, Anne, M. Trueman, C. Evan, S. Chamerro, A. Mireya, P. Pozo, and R. Atkinson

2010 An Extensive Alien Plant Inventory from the Inhabited Areas of Galapagos. *PLOS ONE* 5: 1–8.

Gutiérrez Ortiz, F. J.

2011 Modeling of fire-tube boilers. *Applied Thermal Engineering* 31: 3463–3478.

Habermehl, Karl-Heinz

1961 *Alterbestimmung bei Haustieren, Pelztieren und beim Jagdbaren Wild.* Paul Parey, Berlin.

Hamann, Ole

1981 *Plant Communities of the Galápagos Islands.* Dansk Botanisk Arkiv 34(2), Copenhagen.

1984 Changes and Threats to the Vegetation. In *Galápagos. Key Environments*, edited by R. Perry. Pergamon, Oxford.

2004 Vegetation Changes Over Three Decades on Santa Fe Island, Galápagos, Ecuador. *Nordic Journal of Botany* 23: 143–152.

Hannon, G., Stefan Wastega, Emily Bradshaw, and Richard H. W. Bradshaw

2001 Human Impact and Landscape Degradation on the Faroe Islands. *Biology and Environment: Proceedings of the Royal Irish Academy* 101B: 129–139.

Hazard, Samuel

1871 *Cuba with Pen and Pencil.* Hartford Publishing, Hartford.

Hennessy, Elizabeth

2017 Mythologizing Darwin's Islands. In *Darwin, Darwinism and Conservation in the Galapagos Islands*, edited by D. Quiroga and A. Sevilla, pp. 65–90. Springer International, Cham.

Hennessy, Elizabeth, and Amy L. Cleary

2011 Nature's Eden? The Production and Effects of 'Pristine' Nature in the Galápagos Islands. *Island Studies Journal* 6: 131–156.

Herrera, Adelaida, R. H. Bustamente, and S. A. Shepherd

2012 The Fishery for Endemic Chitons in the Galapagos Islands. *Noticias de Galápagos* 62: 24–28.

Herskovitz, Robert M.

1978 *Fort Bowie Material Culture.* University of Arizona Press, Tucson.

Heyerdahl, Thor, and Arne Skjőlsvold

1956 *Archaeological Evidence of Pre-Spanish Visits to the Galápagos Islands.* Memoirs of the Society for American Archaeology No. 12. Society for American Archaeology, Salt Lake City.

Higham, Charles F. W.

1969 The Metrical Attributes of Two Samples of Bovine Limb Bones. *Journal of Zoology* 157: 63–74.

Higham, Charles, and Michael Message

1970 An Assessment of a Prehistoric Technique of Bovine Husbandry. In *Science in Archaeology: a Survey of Progress and Results*, edited by D. Brothwell and E. Higgs, pp. 315–330. Praeger, New York.

Hlasiwetz, H., and L. Barth

1864 Ueber Einen Neuen, dem Orcin Homologen Kőrper. *Annalen der Chemie* 130: 354–359.

Hobbs, N. Thompson

1996 Modification of Ecosystems by Ungulates. *Journal of Wildlife Management* 60: 695–713.

Hobbs, Richard J., S. Arico, J. Aronson, J. S. Baron, P. Bridgewater, V. A. Cramer, P. R. Epstein, J. J. Ewell, C. A. Klink, A. E. Lugo, D. Norton, D. Ojima, D. M. Richardson, E. W. Sanderson, F. Valladares, M. Vilä, R. Zamora, and M. Zobel

2006 Novel Ecosystems: Theoretical and Management Aspects of the New Ecological World Order. *Global Ecology and Biogeography* 15: 1–7.

Hobbs, Richard J., Lauren M. Hallett, Paul R. Ehrich, and Harold A. Mooney

2011 Intervention Ecology: Applying Ecological Science in the Twenty-first Century. *BioScience* 61: 442–450.

Hobbs, Richard J., Eric S. Higgs, and Carol M. Hall (editors)
2013 *Novel Ecosystems: Intervening in the New Ecological World Order*. Wiley-Blackwell, Chichester.

Hobbs, Richard J., Eric Higgs, Carol M. Hall, Peter Bridgewater, F. Stuart Chapin III, Erle C. Ellis, John J. Ewel, Lauren M. Hallett, James Harris, Kristen B. Hulvey, Stephen T. Jackson, Patricia L. Kennedy, Christoph Kueffer, Lori Lach, Trevor C. Lantz, Ariel E. Lugo, Joseph Mascaro, Stephen D. Murphy, Cara R. Nelson, Michael P. Perring, David M. Richardson, Timothy R. Seastedt, Rachel J. Standish, Brian M. Starzomski, Katherine N. Suding, Pedro M. Tognetti, Laith Yakob, and Laurie Yung
2014 Managing the Whole Landscape: Historical, Hybrid, and Novel Ecosystems. *Frontiers in Ecology and Environment* 12: 557–564.

Hoeck, H. N.
1984 Introduced Fauna. In *Galápagos: Key Environments*, edited by R. Perry, pp. 233–245. Pergamon, Oxford.

Hoff, Stein
1985 *Drømmen om Galapagos: En Ujkent Norsk Utvandrerhistorie* [The Galápagos Dream: An Unknown History of Norwegian Emigration, translated by F. Horneman]. Grødahl and Son, Oslo. http://galapagos.to/TEXTS/HOFF-1.php

Hofman, Courtney, and Torben C. Rick
2018 Ancient Biological Invasions and Island Ecosystems: Tracking Translocations of Wild Plants and Animals. *Journal of Archaeological Research* 26: 65–115.

Holloway, Verity
2016 *The Mighty Healer: Thomas Holloway's Victorian Patent Medicine Empire*. Pen & Sword History, Barnsley, South Yorkshire.

Houze, Herb
2011 *Winchester Repeating Arms Company*. Krause, Iola, Wisconsin.

Hubbell, Mark S. (editor)
1893 *Our Police and Our City: The Official History of the Buffalo Police Department from the Earliest Days to the Present Time*. Bensler and Wesley, Buffalo.

Hunt, Terry L., and Carl P. Lipo
2009 Revisiting Rapa Nui (Easter Island) "Ecocide." *Pacific Science* 63: 601–616.

Hunter, Malcolm L., Kent H. Redford, and David B. Lindenmayer
2014 The Complementary Niches of Anthropocentric and Biocentric Conservationists. *Conservation Biology* 28: 641–645.

Idrovo, Hugo
2005 *Galápagos: Huellas en el Paraiso*. Ediciones Libri Mundi, Quito
2008 *Baltra-Base Beta: Galápagos y la Segunda Guerra Mundial*. Ministerio de Cultura del Ecuador, Quito.

INGALA (Instituto Nacional Galápagos)
2005 *Plan de Manejo del Parque Nacional Galápagos*. Instituto Nacional Galápagos, Presidencia de la República del Ecuador, Quito.

IUCN (International Union for Conservation of Nature)
2018 *The IUCN Red List of Threatened Species. Version 2018-2*. http://www.iucnredlist.org (accessed October 9, 2018).

Jackson, Michael H.

1993 *Galápagos: A Natural History*. University of Calgary, Calgary.

Jaramillo, M., M. Donaghy-Cannon, F. H. Vargas, and P. G. Parker

2016 The Diet of the Galapagos Hawk (*Buteo galapagoensis*) Before and After Goat Eradication. *Journal of Raptor Research* 50: 33–44. https://doi.org/10.3356/rapt-50-01-33-44.1

Jiménez, Ligia M., Susy Mendéz, Susana Donner, Javier Cañón, and Oscar Cortés

2012 Colombian Creole Horse Breed: Same Origin but Different Diversity. *Genetics and Molecular Biology* 35: 790–796.

Jimenez, Roberto

1981 Composition and Distribution of Phytoplankton in the Upwelling System of the Galapagos Islands. In *Coastal Upwelling*, edited by Francis A. Richards, pp. 329–338. American Geophysical Union, Washington, DC.

Jiménez de la Espada, Marcos

1891 Las Islas de los Galápagos y Otras más á Poniente. *Boletín de la Sociedad Geográfica de Madrid* 31: 351–402.

Jones, Allison

2000 Effects of Cattle Grazing on North American Ecosystems: A Quantitative Review. *Western North American Naturalist* 60: 155–164.

Jørgensen, Peter Møller, and Susana León-Yánez

1999 *Catalogue of the Vascular Plants of Ecuador*. Missouri Botanical Garden Press, St. Louis.

Kareiva, Peter, Robert Lalasz, and Michelle Marvier

2011 Conservation in the Anthropocene: Beyond Solitude and Fragility. *Breakthrough Journal* 2: 29–37.

Kareiva, Peter, and Michelle Marvier

2012 What is Conservation Science? *BioScience* 62: 962–969.

Kareiva, Peter, Sean Watts, Robert McDonald, and Tim Boucher

2007 Domesticated Nature: Shaping Landscapes and Ecosystems for Human Welfare. *Science* 316: 1866–1869.

Kastdalen, Alf

1982 Changes in the Biology of Santa Cruz Island between 1935 and 1965. *Noticias de Galápagos* 35: 7–12.

Kearney, Ruth Elizabeth

1935 The Magdalena Bubble. *Pacific Historical Review* 4(1): 25–38.

Kennett, D., Atholl Anderson, Matthew Prebble, Eric Conte, and John Southon

2005 Prehistoric Human Impacts on Rapa, French Polynesia. *Antiquity* 80: 340–354.

Kerr, Suzi, Susan Cardenas, and Joanna Hendy

2004 Migration and the Environment in the Galapagos: An Analysis of Economic and Policy Incentives Driving Migration, Potential Impacts from Migration Control, and Potential Policies to Reduce Migration Pressure. *Motu Working Paper* 03-17. Motu Economic and Public Policy Research, Wellington, NZ.

Key, Gillian, and Edgar Muñoz Heredia
1994 Distribution and Current Status of Rodents in the Galápagos. *Noticias de Galápagos* 53: 21–23.
Kirch, Patrick V.
2000 *On the Road of the Winds: An Archaeological History of the Pacific Islands before European Contact*. University Of California, Berkeley.
2005 Archaeology and Global Change: The Holocene Record. *Annual Review of Environment and Resources* 30: 409–440.
Kowalsky, Arnold A., and Dorothy E. Kowalsky
1999 *Encyclopedia of marks on American, English, and European earthenware, ironstone, and stoneware, 1780–1980*. Atglen, Pennsylvania, Schiffer Publishing.
Kreutzer, L. A.
1992 Bison and Deer Bone Mineral Densities: Comparisons and Implications for the Interpretation of Archaeological Faunas. *Journal of Archaeological Science* 19: 271–294.
Kyselý, René
2016 The Size of Domestic Cattle, Sheep, Goats and Pigs in the Czech Neolithic and Eneolithic Periods: Temporal Variations and their Causes. *Archaeofauna* 25: 33–78.
Lane, Kris
1998 *Pillaging the Empire: Piracy in the Americas 1500–1750*. M. E. Sharpe, Armonk.
Laso, Francisco
2016 Food, Land Use, and Tortoises in the Galapagos. Summer 2016 Field Report. Manuscript on file. Galápagos Academic Institute of Arts and Sciences, Puerto Baquerizo Moreno, Galápagos, Ecuador.
Latimer, Jon
2009 *Buccaneers of the Caribbean: How Piracy Forged an Empire*. Harvard University, Cambridge, MA.
Latorre, Octavio
1991 *Manuel J. Cobos, Emperador de Galápagos*. Fundación Charles Darwin para las Islas Galápagos, Quito.
2002 *Manuel J. Cobos: Su Vida y Su Obra*. Agéncia Española de Cooperación, Quito.
2011 *Historia Humana de Galápagos: Nuevas Investigaciones*. Octavio Latorre, Quito.
2012 *The Curse of the Giant Tortoise: Tragedies, Crimes and Mysteries in the Galápagos Islands*. 7th Edition. Octavio Latorre, Quito.
Lavoie, C., F. Cruz, G. V. Carrion, K. Campbell, C. J. Donlan, S. Harcourt, and M. Moya
2007 *The Thematic Atlas of Project Isabela*. Charles Darwin Foundation, Puerto Ayora, Ecuador.
Lawesson, Jonas E., and Ivan Estupiñan
1987 The Reforestation Project on San Cristobal Island. *Noticias de Galápagos* 45: 23.
Linn, Meredith B.
2008 From Typhus to Tuberculosis and Fractures in Between: A Visceral Historical Archaeology of Irish Immigrant Life in New York City 1845–1870. Unpublished PhD dissertation, Columbia University, New York.

Lugo, Ariel
2009 The Emerging Era of Novel Tropical Forests. *Biotropica* 41: 589–591.
Lundh, Jacob
1999 The Galápagos: A Brief History. Unpublished, Oslo. http://www.lundh.no/jacob/galapagos/pg05.htm
2006a The Last Days of a Paradise. Unpublished, Oslo. http://www.galapagos.to/TEXTS/LUNDH3-1.php
2006b The Farm Area and Cultivated Plants in Santa Cruz, 1932–1965, with Remarks on Other Plants of Galapagos. *Galapagos Research* 64: 12–25.
Lyman, R. Lee
1994 *Vertebrate Taphonomy.* Cambridge University Press, Cambridge.
Lynch, Michelle R.
2015 Heiltsuk Adoption of Euro-American Material Culture at Old Bella Bella, British Columbia, 1833–1899. Unpublished MA thesis, Simon Fraser University, Burnaby, BC.
Lyons, Barry J.
2006 *Remembering the Hacienda: Religion, Authority, and Social Change in Highland Ecuador.* University of Texas Press, Austin.
MacNutt, Francis Augustus
1912[1530] *De Orbo Novo. The Eight Decades of Peter Martyr D'Angheira.* Vol. 1. Burt Franklin, New York.
Mann, Alexander
1909 *Yachting the Pacific. Together with Notes on Travel in Peru, and an Account of the People and Products of Ecuador.* Duckworth and Co., London.
Mann, Daniel, J. Edwards, W. Beck, Richard Reanier, Michele Mass, Bruce Finney, and John Loret
2008 Drought, vegetation change, and human history on Rapa Nui (Isla de Pascua, Easter Island). *Quaternary Research* 69: 16–28.
Markham, Clements
1907[1572] *History of the Incas*, by Pedro Sarmiento de Gamboa. Translated by Sir Clements Markham. Hakluyt Society, Cambridge.
Martin, Paul S., and David W. Steadman
1999 Prehistoric extinctions on islands and continents. In *Extinctions in Near Time: Causes, Contexts and Consequences*, edited by R. D. E. McPhee and H. D. Sues, pp. 17–55. Kluwer, New York.
Martínez, Nicolás
1915 *Impresiones de un Viaje.* Talleres de Policía Nacional, Quito.
Martínez-Abrain, Alejandro, and Juan Jiménez
2016 Anthropogenic Areas as Incidental Substitutes for Original Habitat. *Conservation Biology* 30: 593–598.
Marx, Jennifer
1992 *Pirates and Privateers of the Caribbean.* Krieger, Malabar, FL.

Massertie, F.
1601–1700 Manuscrit Francais No. 385. Département des manuscrits, Bibliothèque nationale de France, Paris. http://gallica.bnf.fr/ark:/12148/btv1b9060640j
Matolcsi, J.
1970 Historische Erforschung der Kőrpergrosse des Rindes auf Grund von Ungarnischen Knochenmaterial. *Zeitschrift für Tierzüchtkunde und Züchtungsbiologie* 87: 89–137.
Matthews, Christopher N., and Allison Manfra McGovern
2018 Created Communities: Segregation and the History of Plural Sites on Eastern Long Island, New York. *Historical Archaeology* 52(1): 30–50.
Mauchamp, A.
1997 Threats from Alien Plant Species in the Galápagos Islands. *Conservation Biology* 11: 260–263.
McMullen, Conley K.
1999 *Flowering Plants of the Galápagos*. Comstock Publishing Associates, Ithaca.
Mellars, Paul
2006 Going East: New Genetic and Archaeological Perspectives on the Modern Human Colonization of Eurasia. *Science* 313: 796–800.
Meyers, Allan Dale
2012 *Outside the Hacienda Walls: The Archaeology of Plantation Peonage in Nineteenth-Century Yucatán*. University of Arizona Press, Tucson.
Meyers, Allan D., and David L. Carlson
2002 Peonage, Power Relations, and the Built Environment at Hacienda Tabi, Yucatan, Mexico. *International Journal of Historical Archaeology* 6(4): 225–252.
Mills, L. S., M. E. Soulée, and D.F. Doak
1993 The Keystone-species-concept in Ecology and Conservation. *Bioscience* 43: 219–224.
Mintz, Sidney W.
1985 *Sweetness and Power*. Penguin, New York.
Moreno Fraginals, Manuel
1976 *The Sugarmill: The Socioeconomic Complex of Sugar in Cuba 1760–1860*. Monthly Review Press, New York.
Moreno Tejada, Jaime
2016 Lazy Labor, Modernization, and Coloniality: Mobile Cultures between the Andes and the Amazon around 1900. *Transfers* 6(2): 4–22.
Morrell, Benjamin, Jr.
1832 *A Narrative of Four Voyages to the South Sea and South Pacific Ocean, Chinese Sea, Ethiopic and Southern Atlantic Ocean, Indian and Antarctic Oceans from the Year 1822 to 1831*. J. and J. Harper, New York.
Morse, Nathaniel B., Paula A. Pellissier, Elisabeth N. Cianciola, Richard L. Brereton, Marleigh M. Sullivan, Nicholas K. Shonka, Tessa B. Wheeler, and William H. McDowell
2014 Novel Ecosystems in the Anthropocene: A revision of the Novel Ecosystem Concept for Pragmatic Applications. *Ecology and Society* 19(2): 12. http://dx.doi.org/10.5751/ES-06192-190212

Mulholland, Susan C.

1989 Phytolith shape frequencies in North Dakota grasses: a comparison to general patterns. *Journal of Archaeological Science* 16: 489–511.

Mulvaney, John, and Johan Kamminga

1999 *Prehistory of Australia*. Smithsonian Institution Press, Washington, DC.

Murcia, Carolina, James Aronson, Gustavo H. Kattan, David Moren-Mateos, Kingsley Dixon, and Daniel Simberloff

2014 A Critique of the 'Novel Ecosystem' Concept. *Trends in Ecology and Evolution* 29: 548–553.

Murillo Posada, Juan Carlos

2010 Population Ecology of Two Species of Chiton *Chiton goodalli* and *Chiton sulcatus*, in the Rocky Coastline of San Cristóbal Island, Galápagos. *Galapagos Report* 2009–2010: 70–75.

Nayton, Gaye

2011 *The Archaeology of Market Capitalism: A Western Australian Perspective.* Springer, New York.

Neira, Cesar Eduardo

2016 Case Study on the Galapagos Islands: Balance for Biodiversity and Migration. *Environmental and Earth Law Journal* 6: 131–152.

Netz, Reviel

2004 *Barbed Wire: An Ecology of Modernity*. Wesleyan University Press, Middletown, CT.

Nicholls, Henry

2013 The 18-km^2 Rat Trap. *Nature* 496: 306–308.

Noddle, Barbara

1973 Determination of the Body Weight of Cattle from Bone Measurements. In *Domestikationsforschung und Geschichte der Hausetiere*, edited by J. Matolcsi, pp. 377–389. Akademiai Kaidó, Budapest.

1974 Age of Epiphyseal Closure in Feral and Domestic Goats and Ages of Dental Eruption. *Journal of Archaeological Science* 1: 195–204.

O'Connell, J. F., and J. Allen

2015 The process, biotic impact, and global implications of the human colonization of Sahul about 47,000 years ago. *Journal of Archaeological Science* 56: 73–84.

O'Connor, Erin

2007 *Gender, Indian, Nation: The Contradictions of Making Ecuador, 1830–1925*. University of Arizona Press, Tucson.

Orlove, Benjamin S.

1997 *The Allure of the Foreign: Imported Goods in Postcolonial Latin America*. University of Michigan Press, Ann Arbor.

Otterman, Lillian

1983 *Clinker Islands: The Mysterious Galapagos.* Great Western, Burbank, CA.

1993 *Clinker Islands: A Complete History of the Galapagos Archipelago*. McGuinn and McGuire, Bradenton.

Pare, Mike
2010 Alstom Site Part of City's Manufacturing DNA. *Chattanooga Times Free Press*, June 2010.
Patton, James L., Suh Y. Yang and Philip Myers
1975 Genetic and Morphologic Divergence among Introduced Rat Populations (*Rattus rattus*) of the Galápagos Archipelago, Ecuador. *Systematic Zoology* 24: 296–310.
Pedrotta, Victoria, and Vanesa Bagaloni
2005 Looking at Interethnic Relations in the Southern Border through Glass Remains: The Nineteenth Century Pampa Region, Argentina. *International Journal of Historical Archaeology* 9(3): 177–193.
Phillips, R. Brand, David A. Wiedenfeld, and Howard L. Snell
2012 Current Status of Alien Vertebrates in the Galápagos Islands: Invasion History, Distribution, and Potential Impacts. *Biological Invasions* 14: 461–480.
Pineo, Ronn F.
1996 *Social and Economic Reform in Ecuador: Life and Work in Guayaquil*. University of Florida Press, Gainesville.
Piperno, Dolores R.
2006 *Phytoliths: A Comprehensive Guide for Archaeologists and Paleoecologists*. AltaMira Press, Lanham.
Piperno, Dolores R., and Deborah M. Pearsall
1998 *The Silica Bodies of Tropical American Grasses: Morphology, Taxonomy, and Implications for Grass Systematics and Fossil Phytolith Identification*. Smithsonian Institution Press, Washington, DC.
Porter, David
1822 *Journal of a Cruise Made to the Pacific Ocean, by Capt. David Porter, in the United States Frigate Essex, In the Years 1812, 1813, and 1814*. Vol. 1, 2nd Edition. Wiley and Halstead, New York.
Porter, Valerie, Laurence Alderson, Stephen J. G. Hall, and Phillip Sponenberg
2016 *Mason's World Encyclopedia of Livestock Breeds and Breeding*. Centre for Agriculture and Biosciences International, Wallingford.
Praetzellis, Mary, Betty J. Rivers, and Jeanette K. Schulz
1983 *Ceramic marks from old Sacramento*. Sacramento, California, California Department of Parks and Recreation.
Pyšek, Petr, David M. Richardson, Marcel Rejmánek, Grady L. Webster, Mark Williamson, and Jan Kirshner
2004 Alien Plants in Checklists and Floras: Towards Better Communication Between Taxonomists and Ecologists. *Taxon* 53: 131–143.
Quiroga, Diego
2009 Crafting Nature: the Galapagos and the Making and Unmaking of a "Natural Laboratory." *Journal of Political Ecology* 16: 123–140.
2013 Changing Views of the Galapagos. In *Science and Conservation in the Galapagos Islands: Frameworks and Perspectives*, edited by S. J. Walsh and C. F. Mena, pp. 23–48. Springer, New York.

Quiroga, Diego, and Gonzalo Rivas

2017 Darwinian Emergence, Conservation, and Restoration. Novel Ecosystems and Hybrid Environments. In *Darwin, Darwinism and Conservation in the Galapagos Islands*, edited by D. Quiroga and A. Sevilla, pp. 151–164. Springer International, Cham.

Quiroga, Diego, and Ana Sevilla

2017 Darwin's Galapagos Myth. In *Darwin, Darwinism and Conservation in the Galapagos Islands*, edited by D. Quiroga and A. Sevilla, pp. 1–7. Springer International, Cham.

Ravines, Rogger

2008 Botellas, frascos y pomos de vidrio en el Perú: un registro inicial. *Boletín de Lima* 30(152): 63–112.

Redclift, M. R.

1978 *Agrarian Reform and Peasant Organization on the Ecuadorian Coast.* Athlone Press, University of London, London.

Reitz, Elizabeth J.

2017 Animal Use at Early Colonies on the Southeastern Coast of the United States. In *The Oxford Handbook of Zooarchaeology*, edited by A. Albarella, M. Rizzetto, H. Russ, K. Vickers, and S. Viner-Daniels, pp. 592–605. Oxford University, Oxford.

Reitz, Elizabeth J., and Barbara Ruff

1994 Morphometric Data for Cattle from North America and the Caribbean Prior to the 1850s. *Journal of Archaeological Science* 21: 699–713.

Reponen, Sini E. M., Sarah K. Brown, Bruce D. Barnett, and Benjamin N. Sacks

2014 Genetic and Morphometric Evidence on a Galápagos Island Exposes Founder Effects and Diversification in the First-known (Truly) Feral Western Dog Population. *Molecular Ecology* 23: 269–283.

Restrepo, A., P. Colinvaux, M. Bush, A. Correa-Metrio, J. Conroy, M. R. Gardener, P. Jaramillo, M. Steinitz-Kannan and J. Overpeck

2012 Impacts of Climate Variability and Human Colonization on the Vegetation of the Galápagos Islands. *Ecology* 93: 1853–1866.

Revidatti, M. A., J. V. Delgado Bermejo, L. T. Gama, V. Landi Periati, C. Ginja, L. A. Alvarez, J. L. Vega-Pla, A. M. Martínez, and BioPig Consortium

2014 Genetic characterization of local Criollo pig breeds from the Americas using microsatellite markers. *Journal of Animal Science* 92: 4823–4832.

Richards, Rhys

1993 Rongotute, Stivers and 'Other Visitors' to New Zealand 'Before Captain Cook.' *Journal of the Polynesian Society* 102: 7–38.

Ritter, Frederick

2015 *Adam and Eve in the Galápagos.* BoD, Norderstedt.

Rodríguez, Linda Alexander

1985 *The Search for Public Policy: Regional Politics and Governmental Finances in Ecuador, 1830–1940.* University of California, Berkeley.

Rouse, John E.

1970 *World Cattle. Vol. 1: Cattle of Europe, South America, Australia, and New Zealand*. University of Oklahoma Press, Norman.

1977 *The Criollo: Spanish Cattle in the Americas*. University of Oklahoma Press, Norman.

Royle, Thomas C. A.

2017 Ancient DNA Analysis of Turtle Remains from El Progreso (CE 1870–1904), San Cristóbal Island, Galápagos Islands, Ecuador. Ms. on file, Ancient DNA Laboratory, Simon Fraser University, Burnaby.

Ruiz, Diego, Angel Chiriboga, and Stuart Banks

2011 CDF Checklist of Galapagos Fish. In *Charles Darwin Foundation Galapagos Species Checklist*, edited by F. Bungartz et al. Charles Darwin Foundation, Puerto Ayora.

Rulau, Russell

2000 *Latin American Tokens: An Illustrated, Priced Catalog of the Unofficial Coinage of Latin America—used in Plantation, Mine, Mill, and Dock—from 1700 to the 20th Century*. Krause, Iola, WI.

Salvador Ayala, Gloria María

2015 *Análisis de Sistema de Producción y Abastecimiento de Alimentos en Galápagos*. Masters thesis, Facultad Latino Americano de Ciencias Sociales Sede Ecuador, Quito.

Salvin, Osbert

1876 On the Avifauna of the Galapagos Archipelago. *Transactions of the Zoological Society of London* 9: 447–510.

Samson, Jane

1999 An Empire of Science: The Voyage of HMS *Herald*, 1845–1851. In *Pacific Empires: Essays in Honour of Glyndwr Williams*, edited by Alan Frost and Jane Samson, pp. 69–85. University of British Columbia Press, Vancouver.

Schávelzon, Daniel

1994 *Arqueología e Historia de la Imprenta Coni, Buenos Aires*. South Carolina Institute of Archaeology and Anthropology, Columbia, SC.

Schelhas, John

2010 The U S National Parks in International Perspective: The Yellowstone Model or Conservation Syncretism? In *National Parks: Vegetation, Wildlife, and Threats*, edited by G. Polisciano and O. Farina, pp. 83–103. NOVA Science, Hauppauge, NY.

Schiller, Laurenne, Juan José Alava, Jack Grove, Günther Reck, and Daniel Pauly

2015 The Demise of Darwin's Fishes: Evidence of Fishing Down and Illegal Shark Finning in the Galápagos Islands. *Aquatic Conservation: Marine and Freshwater Ecosystems* 25: 431–446.

Schodt, David W.

1987 *Ecuador: An Andean Enigma*. Westview Press, Boulder.

Schofield, Eileen K.

1989 Effects of Introduced Plants and Animals on Island Vegetation: Examples from the Galápagos Archipelago. *Conservation Biology* 3: 227–238.

Schulz, Peter D., Betty J. Rivers, Mark M. Hales, Charles A. Litzinger, and Elizabeth A. McKee

1980 *The Bottles of Old Sacramento*. Department of Parks and Recreation, Sacramento, CA.

Schurz, William Lytle

1922 The Spanish Lake. *Hispanic American Historical Review* 5: 181–194.

Seeman, Berthold

1853 *Narrative of the Voyage of H.M.S. Herald During the Years 1845–51 Under the Command of Captain Henrry Kellett, R.N., C.B.; being a Circumnavigation of the Globe and Three Cruises to the Arctic Regions in Search of Sir John Franklin*. Vol. 1. Reeve and Co., London.

Sevilla, Ana

2017a *On The Origin of Species* and the Galapagos Islands. In *Darwin, Darwinism and Conservation in the Galapagos Islands*, edited by D. Quiroga and A. Sevilla, pp. 9–22. Springer International, Cham.

2017b The Galapagos Islands and the Ecuadorian State: Early Encounters. In *Darwin, Darwinism and Conservation in the Galapagos Islands*, edited by D. Quiroga and A. Sevilla, pp. 23–39. Springer International, Cham.

Silver, I. A.

1970 The Ageing of Domestic Animals. In *Science in Archaeology: A Survey of Progress and Results*, edited by D. Brothwell and E. Higgs, pp. 283–302. Praeger, New York.

Sjögren, Guy

2013 The Rise and Decline of the Birmingham Cut-Nail Trade, c. 1811–1914. *Midland History* 38(1): 36–57.

Slevin, Joseph Richard

1931 *Log of the Schooner "Academy" on a Voyage of Scientific Research to the Galapagos Islands 1905–1906*. Occasional Papers Vol. 17. California Academy of Science, San Francisco.

1959 *The Galápagos Islands: A History of their Exploration*. Occasional Papers No. 25. California Academy of Sciences, San Francisco.

Smith, A. G., and A. J. Ferreira

1977 Chiton Fauna of the Galápagos Islands. *Veliger* 20: 82–97.

Smith, C. Lavett

1971 A Revision of the American Groupers: *Epinephelus* and Allied Genera. *Bulletin of the American Museum of Natural History* 146: 67–242.

Snell, Howard, and Solanda Rea

1999 The 1997–98 El Niño in Galápagos: Can 34 Years of Data Estimate 120 Years of Pattern? *Noticias de Galápagos* 60: 11–20.

Soria, M. C., M. R. Gardener, and A. Tye

2002 Eradication of Potentially Invasive Plants with Limited Distributions in the

Galapagos Islands. In *Turning the Tide: The Eradication of Invasive Species*, edited by C. R. Veitch and M. N. Clout, pp. 287–293. Occasional Papers of the IUCN Species Survival Commission 27, Gland.

Soulé, Michael

2013 The "New Conservation." *Conservation Biology* 27: 895–897.

Soulé, Michael E., D. T. Bolger, A. C. Alberts, J. Wright, M. Sorice, and S. Hill

1988 Reconstructed Dynamics of Rapid Extinctions of Chaparral-Requiring Birds in Urban Habitat Islands. *Conservation Biology* 2: 75–92.

Sprague, Roderick

2002 China or Prosser Button Identification and Dating. *Historical Archaeology* 36(2): 111–125.

Spude, Catherine Holder

2005 *The Mascot Saloon: Archeological Investigations in Skagway, Alaska, Vol. 10*. National Park Service, Anchorage.

Squier, George L., and Brother

1870 *The American Sugar Manual: Circular of the American Cane Machinery Manufactured by George L. Squier and Brother, Successors to the Buffalo Agricultural Machine Works, Buffalo N.Y., B.J. West, Agent, New Orleans, La.* George L. Squier and Brother, Buffalo.

1879 *Maquinaria con Privilegio para Haciendas*. George L. Squier and Brother, Buffalo.

Stackpole, Edouard A.

1972 *Whales and Destiny: The Rivalry between America, France, and Britain for Control of the Southern Whale Fishery, 1785–1825*. University of Massachusetts Press, Amherst.

Stahl, Peter W.

2017 Zooarchaeological Data from the Historic Hacienda El Progreso San Cristóbal Island, Galápagos, Ecuador. Open Context. http://opencontext.org/projects/bb308d84-da93-4e0b-8b46-2160c023f592, https://doi.org/10.6078/M7MS3QTG

Standish, Rachel J., Allen Thompson, Eric S. Higgs, and Stephen D. Murphy

2013 Concerns about Novel Ecosystems. In *Novel Ecosystems: Intervening in the New Ecological World Order*, edited by R. J. Hobbs, E. S. Higgs, and Carol M. Hall, pp. 296–309. Wiley-Blackwell, Chichester.

Steadman, David W.

1986 *Holocene Vertebrate Fossils from Isla Floreana, Galápagos*. Smithsonian Contributions to Zoology 413. Smithsonian Institution Press, Washington, DC.

2006 *Extinction and Biogeography of Tropical Pacific Birds*. University of Chicago Press, Chicago.

Steadman, David W., and Clayton E. Ray

1982 *The Relationship of* Megaoryzomys curio, *an Extinct Cricetine Rodent (Muroidea: Muridae) from the Galápagos Islands, Ecuador*. Smithsonian Contributions to Paleobiology 51. Smithsonian Institution Press, Washington, DC.

Steadman, David W., Thomas W. Stafford Jr., Douglas J. Donahue, and A. J. T. Hull
1991 Chronology of Holocene Vertebrate Extinction in the Galápagos Islands. *Quaternary Research* 36: 126–133.
Stein, Stanley J., and Barbara H. Stein
2000 *Silver, Trade, and War: Spain and America in the Making of Early Modern Europe*. Johns Hopkins University Press, Baltimore.
Stewart, Alban
1911 *A Botanical Survey of the Galapagos Islands*. Vol. 1. Proceedings of the California Academy of Sciences, San Francisco.
Stoops, Georges
2014 Soils and Paleosols of the Galápagos Islands: What We Know and What We Don't Know, A Meta-Analysis. *Pacific Science* 68: 1–17.
Strauch, Dore
1936 *Satan Comes to Eden, as Told by Dore Strauch to Walter Brodeman*. Harper and Brothers, New York.
Strömberg, Caroline A. E.
2009 Methodological Concerns for Analysis of Phytolith Assemblages: Does Count Size Matter? *Quaternary International* 193(1–2): 124–140.
Strömberg, Caroline A. E., Verónica S. Di Stilio, and Zhaoliang Song
2016 Functions of Phytoliths in Vascular Plants: An Evolutionary Perspective. *Functional Ecology* 30: 1286–1297.
Sullivan, Catherine
1994 Searching for Nineteenth-Century Florida Water Bottles. *Historical Archaeology* 28(1): 78–98.
Sulloway, Frank J.
1982a Darwin and His Finches: The Evolution of a Legend. *Journal of the History of Biology* 15: 1–53.
1982b Darwin's Conversion: The Beagle Voyage and its Aftermath. *Journal of the History of Biology* 15: 325–396.
Sweitz, Sam R.
2012 *On the Periphery of the Periphery: Household Archaeology at Hacienda San Juan Bautista Tabi, Yucatán, Mexico.* Springer-Verlag, New York.
Switzer, Ronald R.
1974 *The Bertrand Bottles: A Study of 19th Century Glass and Ceramic Containers.* National Parks Sevice, Washington, D.C.
Taggart-Hodge, Tanya
2016 *A Century of Landscape-Level Changes in the Bow Watershed, Alberta, Canada, and Implications for Flood Management*. Unpublished Master's thesis, School of Environmental Studies, University of Victoria, Victoria.
Teichert, M.
1975 Osteologische Untersuchungen zur Berechnung der Widerristhőhe bei Schafen. In *Archaeozoological Studies*, edited by A. T. Clason, pp. 51–69. Elsevier, Amsterdam.

Tellkamp, Markus P.

2017 The Galápagos Islands: Relatively Untouched, but Increasingly Endangered. *Bionatura* 2. https://doi.org/10.21931/RB/2017.02.02.2

Tezanos Toral, Lorena

2015 *The Architecture of Nineteenth-Century Cuban Sugar Mills: Creole Power and African Resistance in Late Colonial Cuba*. CUNY Academic Works http://academicworks.cuny/gc_cfds/1156

Tirira, Diego

2007 *Mamíferos del Ecuador: Guía del Campo*. Ediciones Murciélago Blanco, Quito.

Toral-Granda, M. V., C. E. Causton, H. Jäger, M. Trueman, J. C. Araujo E., M. Cruz, K. K. Zander, A. Izurieta, and S. T. Garnett

2017 Alien Species Pathways to the Galapagos Islands, Ecuador. *PLOS ONE* 12(9): e0184379 https://doi.org/10.1371/journal.pone.0184379

Tower, Walter S.

1907 *A History of the American Whale Fishery*. Series in Political Economy and Public Law, No. 20. University of Pennsylvania, Philadelphia.

Townsend, Charles Haskins

1925 The Galapagos Tortoises in Their Relation to the Whaling Industry: A Study of Old Log Books. *Zoologica* 4: 55–135.

Trueman, Mandy, and Noémi d'Ozouville

2010 Characterizing the Galapagos Terrestrial Climate in the Face of Global Climate Change. *Galapagos Research* 67: 26–37.

Trueman, Mandy, Richard J. Hobbs, and Limberly Van Niel

2013 Interdisciplinary Historical Vegetation Mapping for Ecological Restoration in Galápagos. *Landscape Ecology* 2013: 519–532.

Tye, Alan

2001 Invasive Plant Problems and Requirements for Weed Risk Assessment in the Galapagos Islands. In *Weed Risk Assessment*, edited by R. H. Groves, F. D. Panetta, and J. G. Virtue, pp. 135–175. CSIRO, Victoria.

2006 Can We Infer Island Introduction and Naturalization Rates from Inventory Data? Evidence from Introduced Plants in Galapagos. *Biological Invasions* 8: 201–215.

Tye, A., and B. Lau

2014 *Blutaparon rigidum*. *The IUCN Red List of Threatened Species* 2014: e.T39087A61481721. http://dx.doi.org/10.2305/IUCN.UK.2014-3.RLTS.T39087A61481721.en (accessed October 18, 2018).

Tye, Alan., M. C. Soria, and M. R. Gardener

2002 A Strategy for Galapagos Weeds. In *Turning the Tide: The Eradication of Invasive Species*, edited by C. R. Veitch and M. N. Clout, pp. 336–341. IUCN SSC Invasive Species Specialist Group. International Union for Conservation of Nature, Gland.

Utely, Temple

1938 *A Modern Sea Beggar*. Peter Davies, London.

Valle, Carlos A., and Patricia G. Parker
2012 Research on Evolutionary Principles in Galápagos. In *The Role of Science for Conservation*, edited by M. Wolff and M. Gardener, pp. 16–34. Routledge, London.

van der Knaap, W. O., Jacqueline F. N. van Leeuwen, Cynthia A. Froyd, and Kathy J. Willis
2012 Detecting the Provenance of Galápagos Non-Native Pollen: The Role of Humans and Air Currents as Transport Mechanisms. *Holocene* 22: 1373–1383.

Vanegas León, Brenda
1998 *Leyendas y Tradiciones de Galápagos, en la Cultura y Literatura como Expresión Popular.* Casa de La Cultura Ecuatoriana "Benjamín Carrión, Quito.

van Leeuwen, Jacqueline F. N., Cynthia A. Froyd, W. O. van der Knaap, Emily E. Coffey, Alan Tye, and Katherine J. Willis
2008 Fossil Pollen as a Guide to Conservation in the Galápagos. *Science* 322(5905): 1206.

VanValkenburgh, Parker, Sarah J. Kelloway, Karen L. Privat, Bill Sillar, and Jeffrey Quilter
2017 Rethinking Cultural Hybridity and Technology Transfer: SEM Microstructural Analysis of Lead Glazed Ceramics from Early Colonial Peru. *Journal of Archaeological Science* 82: 17–30.

Vargas, Pablo, R. Heleno, A. Traveset, and M. Nogales
2012 Colonization of the Galapagos Islands by Plants with No Specific Syndromes for Long-Distance Dispersal: A New Perspective. *Ecography* 35: 33–43.

Vargas, S., A. Larbi, and M. Sánchez
2007 Analysis of Size and Conformation of Native Creole Goat Breed and Crossbreds Used in Smallholder Agrosilvopastoral Systems in Puebla, Mexico. *Tropical Animal Health and Production* 39: 279–286.

Vecchi, G. A., A. Clement, and B. J. Soden
2008 Examining the Tropical Pacific's Response to Global Warming. *EOS Transactions, American Geophysical Union* 89(9): 81, 83.

Vésteinsson, Orri, and Thomas H. McGovern
2012 The Peopling of Iceland. *Norwegian Archaeological Review* 45: 206–218.

Vidal Gormaz, Francisco
1891 El Archipiélago de las Galápagos. *Boletín de la Sociedad de Madrid* 31: 153–199.

Villa, Angel, and Pool Segarra
2012 Changes in Land Use and Vegetative Cover in the Rural Areas of Santa Cruz and San Cristóbal. *Galapagos Report* 2009–2010: 83–91.

Vince, Gaia
2011 Embracing Invasives. *Science* 331: 1383–1384.

Von Hagen, Victor
1949 *Ecuador and the Galápagos Islands.* University of Oklahoma, Norman.

Wafer, Lionel
1699 *A New Voyage and Description of the Isthmus of America.* James Knapton, London.

Wallis-Tayler, A. J.
1895 *Sugar Machinery. A Descriptive Treatise Devoted to the Machinery Used in the Manufacture of Cane and Beet Sugars.* William Rider and Son, London.
Walter, John
2006 *The Guns that Won the West: Firearms on the American Frontier, 1848–1898.* Greenhill, London.
Ware, Lewis S.
1880 *The Sugar Beet: Including a History of the Beet Sugar Industry in Europe, Varieties of the Sugar Beet, Examination, Soils, Tillage, Seeds and Sowing, Yield and Cost of Cultivation, Harvesting, Transportation, Conservation, Feeding Qualities of the Beet and of the Pulp, etc.* Henry Carey Baird and Company, Philadelphia.
Waters, Michael R., and Thomas Wier Stafford, Jr.
2012 The First Americans: A Review of the Evidence for the Late-Pleistocene Peopling of the Americas. In *Paleoamerican Odyssey*, edited by K. E. Graf, C. V. Ketron, and M. R. Waters, pp. 543–562. Texas A&M University, College Station.
Watkins, Graham, and Felipe Cruz
2007 *Galápagos at Risk: A Socioeconomic Analysis.* Charles Darwin Foundation, Puerto Ayora, Ecuador.
Watson, James, Mandy Trueman, Marta Tufet, Scott Henderson, and Rachel Atkinson
2010 Mapping Terrestrial Anthropogenic Degradation on the Inhabited Islands of the Galapagos Archipelago. *Oryx* 44: 79–82.
Webb, Robert H., Raymond M. Turner, and Diane E. Boyer
2010 Introduction: A Brief History of Repeat Photography. In *Repeat Photography: Methods and Applications in the Natural Sciences*, edited by R. H. Webb, R. M. Turner, and D. E. Boyer, pp. 3–11. Island Press: Washington, DC.
Webster, E. Alfonso
1904 *Los Crímenes de Galápagos (Archipiélago de Colón).* Imprenta de el Telégrafo, Guayaquil.
Whittaker, Robert J., and José María Fernández-Palacios
2007 *Island Biogeography. Ecology, Evolution and Conservation.* 2nd Edition. Oxford University, Oxford.
Wiggins, Ira L., and Duncan M. Porter
1971 *Flora of the Galápagos Islands.* Stanford University, Stanford.
Williams, Derek
2005 The Making of Ecuador's Pueblo Catolico, 1861–1875. In *Political Cultures in the Andes, 1750–1950*, edited by Nils Jacobsen and Cristóbal Aljovín de Losada. Duke University Press, Durham.
Wittmer, Margret
1961 *Floreana Adventure.* E.P. Dutton, New York.
Wolf, Teodoro
1887 *Memoria Sobre Las Islas de Galápagos.* Imprenta del Gobierno, Quito.
Woram, John
1989 Galápagos Island Names. *Noticias de Galápagos* 48: 22–32.

2005 *Charles Darwin Slept Here: Tales of Human History at World's End*. Rockville Press, Rockville Centre, NY.

2016 Las Encantadas: Human and Cartographic History of the Galápagos Islands. http://www.galapagos.to

Zavaleta, Erika S., Richard J. Hobbs, and Harold A. Mooney

2001 Viewing Invasive Species Removal in a Whole-System Context. *Trends in Ecology and Evolution* 16: 454–459.

Zumwalt, Betty

1980 *Ketchup, Pickles, and Sauces: 19th Century Food in Glass*. Mark West Publishers, Fulton, CA.

INDEX

Page numbers followed by the letters *f* and *t* refer to figures and tables.

Peter W. Stahl is professor emeritus at Binghamton University and adjunct professor at the University of Victoria. He is the author of numerous publications on South American archaeology, historical ecology, zooarchaeology, and vertebrate taphonomy, including a recent chapter in *The Oxford Handbook of Zooarchaeology*.

Fernando J. Astudillo is assistant professor of archaeology at the Universidad San Francisco de Quito. He has written articles in *Vegetation History and Archaeobotany* and *Palaeogeography, Palaeoclimatology, Palaeoecology*.

Ross W. Jamieson is associate professor of archaeology at Simon Fraser University. He is the author of *Domestic Architecture and Power: The Historical Archaeology of Colonial Ecuador*.

Diego Quiroga is vice president of research in the Universidad San Francisco de Quito and codirector of the Galápagos Academic Institute of Arts and Sciences (GAIAS). He is the coeditor of *Darwin, Darwinism and Conservation in the Galapagos Islands*.

Florencio Delgado is professor of anthropology and director of CIS at the Universidad San Francisco de Quito. He is the author of many chapters in archaeology books, one of which is *La Arqueología y la Antropología en Ecuador: Escenarios, Retos y Perspectivas* (The Archaeology and Anthropology of Ecuador: Scenarios, Challenges and Perspectives).

SOCIETY AND ECOLOGY IN ISLAND AND COASTAL ARCHAEOLOGY

Edited by Victor D. Thompson

The settlement and occupation of islands, coastlines, and archipelagoes can be traced deep into the human past. From the voyaging and seafaring peoples of the Oceania to the Mesolithic fisher-hunter-gatherers of coastal Ireland, to coastal salt production among Maya traders, the range of variation found in these societies over time is boundless. Yet, they share a commonality that links them all together—their dependence upon seas, coasts, and estuaries for life and prosperity. Thus, in all these cultures there is a fundamental link between society and the ecology of islands and coasts. Books in this series explore the nature of humanity's relationship to these environments from a global perspective. Topics in this series would range from edited volumes to single case studies covering the archaeology of initial migrations, seafaring, insularity, trade, societal complexity and collapse, early village life, aquaculture, and historical ecology, among others along islands and coasts.

The Powhatan Landscape: An Archaeological History of the Algonquian Chesapeake, by Martin D. Gallivan (2016; first paperback edition, 2018)

An Archaeology of Abundance: Reevaluating the Marginality of California's Islands, edited by Kristina M. Gill, Mikael Fauvelle, and Jon M. Erlandson (2019)

Maritime Communities of the Ancient Andes, edited by Gabriel Prieto and Daniel H. Sandweiss (2019)

The Archaeology of Human-Environmental Dynamics on the North American Atlantic Coast, edited by Leslie Reeder-Myers, John A. Turck, and Torben Rick (2019)

Historical Ecology and Archaeology in the Galápagos Islands: A Legacy of Human Occupation, by Peter W. Stahl, Fernando J. Astudillo, Ross W. Jamieson, Diego Quiroga, and Florencio Delgado (2020)